Undergraduate Topics in Computer Science

Undergraduate Topics in Computer Science' (UTiCS) delivers high-quality instructional content for undergraduates studying in all areas of computing and information science. From core foundational and theoretical material to final-year topics and applications, UTiCS books take a fresh, concise, and modern approach and are ideal for self-study or for a one- or two-semester course. The texts are all authored by established experts in their fields, reviewed by an international advisory board, and contain numerous examples and problems. Many include fully worked solutions.

Also in this series

Iain D. Craig
Object-Oriented Programming Languages: Interpretation
978-1-84628-773-2

Max Bramer
Principles of Data Mining
978-1-84628-765-7

Hanne Riis Nielson and Flemming Nielson
Semantics with Applications: An Appetizer
978-1-84628-691-9

Michael Kifer and Scott A. Smolka
Introduction to Operating System Design and Implementation: The OSP 2 Approach
978-1-84628-842-5

Phil Brooke and Richard Paige
Practical Distributed Processing
978-1-84628-840-1

Frank Klawonn
Computer Graphics with Java
978-1-84628-847-0

David Salomon
A Concise Introduction to Data Compression
978-1-84800-071-1

David Makinson
Sets, Logic and Maths for Computing
978-1-84628-844-9

Alan P. Parkes

A Concise Introduction to Languages and Machines

 Springer

Alan P. Parkes, PhD
Honorary Senior Lecturer, Computing Dept., Lancaster University, UK

Undergraduate Topics in Computer Science ISSN 1863-7310
ISBN 978-1-84800-120-6   e-ISBN 978-1-84800-121-3
DOI 10.1007/978-1-84800-121-3

British Library Cataloguing in Publication Data
A catalogue record for this book is available from the British Library

Library of Congress Control Number: 0000000

Printed on acid-free paper

9 8 7 6 5 4 3 2 1

Springer Science+Business Media
springer.com

Contents

Preface

Aims and Objectives

This book focuses on key theoretical topics of computing, in particular formal languages and abstract machines. It is intended primarily to support the theoretical modules on a computer science or computing-related undergraduate degree scheme.

Though the book is primarily theoretical in nature, it attempts to avoid the overly mathematical approach of many books on the subject and for the most part focuses on encouraging the reader to gain an intuitive understanding. Proofs are often only sketched and, in many cases, supported by diagrams. Wherever possible, the book links the theory to practical considerations, in particular the implications for programming, computation and problem solving.

Organisation and Features of the Book

There is a short introductory chapter that provides an overview of the book and its main features. The remainder of the book is in two parts, *Languages and Machines* and *Machines and Computation*.

Part 1, Languages and Machines, is concerned with formal language theory as it applies to Computer Science. It begins with an introduction to the notation and concepts that support the theory, such as strings, the various ways in which a formal language can be defined, and the Chomsky hierarchy of formal language. It then focuses on the languages of the Chomsky hierarchy, in each case also introducing the abstract machines associated with each type of language.

The topics are regular languages and finite state recognisers, context free languages and pushdown recognisers, and context sensitive and unrestricted languages and the Turing machine. Many important theoretical properties of regular and context free languages are established. The more intellectually demanding of these results are mostly confined to a single chapter, so that the reader can focus on the general thrust of the argument of Part 1, which is to demonstrate that the Chomsky hierarchy is indeed a proper hierarchy for both languages and machines, and to consider some of the implications of this.

In the first part of the book the finite state machine, the pushdown machine and the Turing machine are considered as language recognisers, though many hints are given about the potential computational properties of these abstract machines.

Part 2, Machines and Computation, considers the computational properties of the machines from Part 1 in more detail. The relationship between finite state machines and the digital computer is explored. This leads us on to the need for a more powerful machine to deal with arbitrary computation. This machine is shown to be the Turing machine, introduced in Part 1 as a language processor.

The Turing machine is used to explore key aspects of computation, such as non-determinism, parallel processing and the efficiency of computational processes. The latter is considered in the context of a brief introduction to algorithm analysis, using big O notation.

The book also considers the limitations of computation, both in terms of language processing (simply defined formal languages that cannot be processed by any machine) and computation (the halting problem and related issues).

The book contains numerous illustrative figures, and proofs are often partly accomplished through diagrammatic means.

From Chapter 2 onwards, each chapter concludes with exercises, some of which are programming exercises. Solutions and hints for many of the exercises appear at the end of the book.

The book also contains a list of recommended reading material.

For Students

I wrote this book partly because when I studied this material as part of my own Computing degree, I had to work really hard to understand the material, a situation which arose not because the material is too difficult, but because it was not well presented and the books seemed to assume I was a pure mathematician, which I am not.

This book is primarily for undergraduate students of computing, though it can also be used by students of computational linguistics and researchers, particularly those entering computer science from other disciplines, who find that they require a foundation or a refresher course in the theoretical aspects of computing.

Some aspects of the book are certainly clearer if the student has some experience of programming, though such experience is not essential for understanding most of the book.

The reader is advised where especially demanding material can be omitted, though he or she is encouraged to appreciate the implications of that material, as such an appreciation may be assumed later in the book.

For Instructors

I wrote this book partly because when I taught this material to undergraduates, as a Computer Science lecturer in a UK university, I had to work really hard to present the material in an appropriate way, and at the appropriate level, a situation which arose mainly because the available books seemed to assume the students were pure mathematicians, which mostly they are not.

This is intended to be not only a book to promote understanding of the topics for students, but also a recommendation of the core material that should be covered in a theoretical computer science module for undergraduates. I suggest that to cover all of the material would require at least a module in each semester of one year of the degree course. If possible, the linguistic aspects should precede a compiler course and the computational aspects should precede an algorithm analysis course. If the student has programming experience the material is much more accessible. This should influence the way it is presented if it appears early on in the degree scheme.

There is, of course, much more theory relating to programming language design, compiler writing, parallel processing and advanced algorithm analysis than presented in this book. However, such additional theory is best covered in the modules that concern these subjects, some of which are often optional in a computing degree scheme.

1
Introduction

1.1 Overview

This chapter briefly describes:

- what this book is about
- what this book tries to do
- what this book tries not to do
- a useful feature of the book: the exercises.

1.2 What this Book is About

This book is about two key topics of computer science, namely *computable languages* and *abstract machines*.

Computable languages are related to what are usually known as "formal languages". I avoid using the latter phrase here because later on in the book I distinguish between *formal* languages and *computable* languages. In fact, computable languages are a special type of formal languages that can be processed, in ways considered in this book, by computers, or rather *abstract machines* that *represent* computers.

Abstract machines are formal computing devices that we use to investigate properties of *real* computing devices. The term that is sometimes used to describe

A.P. Parkes, *A Concise Introduction to Languages and Machines*,
DOI: 10.1007/978-1-84800-121-3_1, © Springer-Verlag London Limited 2008

abstract machines is *automata*, but that sounds too much like real machines, in particular the type of machines we call *robots*.

This book assumes that you are a layperson, in terms of computer *science*. If you are a computer science undergraduate, as you might be if you are reading this book, you may by now have written many programs. So, in this introduction we will draw on your experience of programming to illustrate some of the issues related to formal languages that are introduced in this book.

The programs that you write are written in a formal language. They are expressed in text which is presented as input to another program, a *compiler* or perhaps an *interpreter*. To the compiler, your program text represents a *code*, which the compiler knows exactly how to handle, as the rules by which that code is constructed are completely specified inside the compiler. The type of language in which we write our programs (the *programming language*), has such a well-defined syntax (rules for forming acceptable programs) that a machine can decide whether or not the program you enter has the right even to be considered as a proper program. This is only part of the task of a compiler, but it's the part in which we are most interested.

For whoever wrote the compiler, and whoever uses it, it is very important that the compiler does its job properly in terms of deciding that your program is syntactically valid. The compiler writer would also like to be sure that the compiler will always *reject* syntactically *in*valid programs as well as accepting those that are syntactically valid. Finally, the compiler writer would like to know that his or her compiler is not wasteful in terms of precious resources: if the compiler is more complex than it needs to be, if it carries out many tasks that are actually unnecessary, and so on. In this book we see that the solutions to such problems depend on the type of language being considered.

Now, because your program is written in a formal language that is such that another program can decide if your program *is* a program, the programming language is a computable language. A *computable* language then, is a *formal* language that is such that a computer can understand its syntax. Note that we have not discussed what your program will actually *do*, when it's run: the compiler doesn't really understand that at all. The compiler is just as happy to compile a program that doesn't do what you intended as it is to compile one that does (as you'll know, if you've ever done any programming).

The book is in two parts. Part 1: Languages and Machines is concerned with the relationship between different types of formal languages and different types of theoretical machines (abstract machines) that can serve to process those languages, in ways we consider later. So, the book isn't really *directly* concerned with *programming* languages. However, much of the material in the first part of the book is highly relevant to programming languages, especially in terms of providing a useful theoretical background for compiler writing, and even programming language design. For that reason, some of the examples used in the book are from programming languages (particularly Pascal in fact, but you need not be familiar with the language to appreciate the examples).

In Part 1, we study four different types of formal languages and see that each of these types of formal language is associated with a particular type of abstract machine. The four types of language we consider were actually defined by the American linguist Noam Chomsky; the classification of formal languages he defined has come to be called the *Chomsky hierarchy*. It *is* a hierarchy, since it defines a general type of language (type 0), then a restricted version of that general type (type 1), then a restricted version of *that* type (type 2), and then finally the most restricted type of all (type 3).

The types of language considered are very simple to define, and even the most complex of abstract machines is also quite simple. What's more, all of the types of abstract machine we consider in this book can be represented in diagrammatic form. The most complex abstract machine we look at is called the Turing machine (TM), after Alan Turing, the mathematician who designed it. The TM is, in some senses, the most powerful computational device possible, as you will see in Part 2 of the book.

The chapters of Part 1 are shown in Figure 1.1.

By the end of Part 1 of the book you should have an intuitive appreciation of computable languages. Part 2: Machines and Computation investigates computation in a wider sense. We see that we can discuss the types of computation carried out by *real* computers in terms of our abstract machines. We see that a machine, called a finite state transducer, which appears to share some properties with real computers, is not really a suitable general model of real computers. Part of the reason for this is that this machine cannot do multiplication or division. We see that the *Turing machine* is capable of multiplication and division, and any other tasks that real computers can do. We even see that the TM can run programs. What is more, since the TM effectively has unlimited storage capacity, it turns out to be *more* powerful than any real computer could ever be.

One interesting property of TMs is that we cannot make them any more powerful than they are by adding extra computational facilities. Nevertheless, TMs use only one data structure, a potentially infinite one-dimensional array of symbols, called a tape, and only one type of instruction. A TM can simply read a symbol from its tape, replace that symbol by another, and then move to the next symbol on the right or left in the tape. We find that if a TM is designed so that it carries out many processes simultaneously, or uses many additional tapes, it cannot perform any more computational tasks than the basic serial one-tape version. In fact, it has been established that no other formalisms we might create

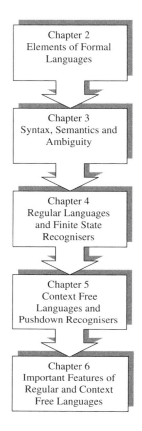

Figure 1.1 The chapters of Part 1: Languages and Machines.

for representing computation give us any more functional power than does a TM. This assertion is known as *Turing's thesis*.

We see that we can actually take any TM, code it and its data structure (tape) as a sequence of zeros and ones, and then let another TM run the original TM as if it were a program. This TM can carry out the computation described by any TM. We thus call it the universal Turing machine (UTM). However, it is simply a standard TM, which, because of the coding scheme used, only needs to expect either a zero or a one every time it examines a symbol on its tape. The UTM is an abstract counterpart of the real computer.

The UTM has unlimited storage, so no real computer can exceed its power. We use this fact as the basis of some extremely important results of computer science, one of which is to see that the following problem cannot be solved, in the general case:

Given any program, and appropriate input values to that program, will that program terminate when run on that input?

In terms of TMs, rather than programs, this problem is known as the *halting problem*. We see that the halting problem is unsolvable,[1] which has implications both for the real world of computing and for the nature of formal languages. We also see that the halting problem can be used to convince us that there are formal languages that cannot be processed by any TM, and thus by any program. This enables us to finally define the relationship between computable languages and abstract machines.

At the end of Part 2, our discussion about abstract machines leads us on to a topic of computer science, algorithm complexity, that is very relevant to

Figure 1.2 The chapters of Part 2: Machines and Computation.

---

[1] You have probably realised that the halting problem is solvable in certain cases.

programming. In particular, we discuss techniques that enable us to predict the running time of algorithms, and we see that dramatic savings in running time can be made by certain cleverly defined algorithms.

Figure 1.2 shows the chapters of Part 2 of the book.

1.3 What this Book Tries to Do

This book attempts to present formal computer science in a way that you, the student, can understand. This book does not assume that you are a mathematician. It assumes that you are not particularly familiar with formal notation, set theory, and so on. As far as is possible, excessive formal notation is avoided. When formal notation or jargon is unavoidable, the formal definitions are usually accompanied by short explanations, at least for the first few times they appear.

Overall, this book represents an understanding of formal languages and abstract machines which lies somewhere beyond that of a layperson, but is considerably less than that of a mathematician. There is a certain core of material in the subject that any student of computer science, or related disciplines, should be aware of. Nevertheless, many students do not become aware of this important and often useful material. Sometimes they are discouraged by the way it is presented. Sometimes, books and lecturers try to cover too much material, and the fundamentals get lost along the way.

Above all, this book tries to highlight the connections between what might initially appear to be distinct topics within computer science.

1.4 What this Book Tries Not to Do

This book tries not to be too formal. References to a small number of formal books are included in the section "Further Reading", at the end of the book. In these books you will find many theorems, usually proved conclusively by logical proof techniques. The "proofs" appearing in this book are included usually to establish something absolutely central, or that we need to assume for subsequent discussion. Where possible, such "proofs" are presented in an intuitive way.

This is not to say that proofs are unimportant. I assume that, like me, you are the type of person who has to convince themselves that something is right before you really accept it. However, the proofs in this book are presented in a different way from the way that proofs are usually presented. Many results are established "beyond reasonable doubt" by presenting particular examples and encouraging the reader to appreciate ways in which the examples can be generalised. Of course, this is not really sound logic, as it does not argue throughout in terms of the general case. Some of the end-of-chapter exercises present an opportunity to practise or complete proofs. Such exercises often include hints and/or sample answers.

1.5 The Exercises

At the end of each of Chapters 2 to 12, you will find a small number of exercises to test and develop your knowledge of the material covered in that chapter. Most of the exercises are of the "pencil and paper" type, though some of them are medium-scale programming problems. Any exercise marked with a dagger (†) has a sample solution in the "Solutions to Selected Exercises" section near the end of the book. You do not need to attempt any of the exercises to fully understand the book. However, although the book attempts to make the subject matter as informal as possible, in one very important respect it is very much like maths: *you need to practise applying the knowledge and skills you learn or you do not retain them.*

Finally, some of the exercises give you an opportunity to investigate additional material that is not covered in the chapters themselves.

1.6 Further Reading

A small section called "Further Reading" appears towards the end of the book. This is not meant to be an exhaustive list of reading material. There are many other books on formal computer science than are cited here. The further reading list also refers to books concerned with other fields of computer science (e.g. computer networks) where certain of the formal techniques in this book have been applied. Brief notes accompany each title cited.

1.7 Some Advice

Most of the material in this book is very straightforward, though some requires a little thought the first time it is encountered. Students of limited formal mathematical ability should find most of the subject matter of the book reasonably accessible. You should use the opportunity to practise provided by the exercises, if possible. If you find a section really difficult, ignore it and go on to the next. You will probably find that an appreciation of the overall result of a section will enable you to follow the subsequent material. Sections you omit on first reading may become more comprehensible when studied again later.

You should not allow yourself to be put off if you cannot see immediate applications of the subject matter. There have been many applications of formal languages and abstract machines in computing and related disciplines, some of which are referred to by books in the "Further Reading" section.

This book should be interesting and relevant to any intelligent reader who has an interest in computer science and approaches the subject matter with an open mind. Such a reader may then see the subject of languages and machines as an explanation of the simple yet powerful and profound abstract computational processes beneath the surface of the digital computer.

Part 1
Languages and Machines

2
Elements of Formal Languages

2.1 Overview

In this chapter, we discuss:

- the building blocks of formal languages: *alphabets* and *strings*

- *grammars* and *languages*

- a way of classifying grammars and languages: the *Chomsky hierarchy*

- how formal languages relate to the definition of programming languages

and introduce:

- writing definitions of sets of strings

- producing sentences from grammars

- using the notation of formal languages.

2.2 Alphabets

An alphabet is a finite collection (or set) of symbols. The symbols in the alphabet are entities which cannot be taken apart in any meaningful way, a property which leads to them being sometimes referred to as *atomic*. The symbols of an alphabet are simply the "characters", from which we build our "words". As already said, an

A.P. Parkes, *A Concise Introduction to Languages and Machines*,
DOI: 10.1007/978-1-84800-121-3_2, © Springer-Verlag London Limited 2008

alphabet is *finite*. That means we could define a program that would print out its elements (or members) one by one, and (this last part is very important) the program would terminate sometime, having printed out each and every element.

For example, the small letters you use to form words of your own language (e.g. English) could be regarded as an alphabet, in the formal sense, if written down as follows:

$$\{a, b, c, d, e, ..., x, y, z\}.$$

The digits of the (base 10) number system we use can also be presented as an alphabet:

$$\{0, 1, 2, 3, 4, 5, 6, 7, 8, 9\}.$$

2.3 Strings

A string is a finite sequence of zero or more symbols taken from a formal alphabet. We write down strings just as we write the words of this sentence, so the word "strings" itself could be regarded as a *string* taken from the alphabet of letters, above. Mathematicians sometimes say that a string taken from a given alphabet is a string *over* that alphabet, but we will say that the string is *taken from* the alphabet. Let us consider some more examples. The string *abc* is one of the many strings which can be taken from the alphabet { *a*, *b*, *c*, *d*}. So is *aabacab*. Note that duplicate symbols are allowed in strings (unlike in sets). If there are no symbols in a string it is called the *empty string*, and we write it as ε (the Greek letter *epsilon*), though some write it as λ (the Greek letter *lambda*).

2.3.1   Functions that Apply to Strings

We now know enough about strings to describe some important functions that we can use to manipulate strings or obtain information about them. Table 2.1 shows the basic string operations (note that x and y stand for *any* strings).

You may have noticed that strings have certain features in common with *arrays* in programming languages such as Pascal, in that we can index them. To index a string, we use the notation x_i, as opposed to something like $x[i]$. However, strings actually have more in common with the list data structures of programming languages such as LISP or PROLOG, in that we can concatenate two strings

Table 2.1 The basic operations on strings.

Operation	Written as	Meaning	Examples and comments
length	$\|x\|$	the number of symbols in the string x	$\|abcabca\| = 7$ $\|a\| = 1$ $\|\varepsilon\| = 0$
concatenation	xy	the string formed by writing down the string x followed immediately by the string y concatenating the empty string to any string makes no difference	let $x = abca$ let $y = ca$ then: $\quad xy = abcaca$ let $x = \,<any\ string>$ then: $\quad x\varepsilon = x$ $\quad \varepsilon x = x$
power	x^n, where n is a whole number ≥ 0	the string formed by writing down n copies of the string x	let $x = abca$ then: $\quad x^3 = abcaabcaabca$ $\quad x^1 = x$ Note: $\quad x^0 = \varepsilon$
index	x_i, where i is a whole number	the i th symbol in the string x (i.e. treats the string as if it were an array of symbols)	let $x = abca$ then: $\quad x_1 = a$ $\quad x_2 = b$ $\quad x_3 = c$ $\quad x_4 = a$

together, creating a new string. This is like the *append* function in LISP, with strings corresponding to lists, and the empty string corresponding to the empty list. It is only possible to perform such operations on arrays if the programming language allows arrays to be of dynamic size. (which Pascal, for example, does not). However, many versions of Pascal now provide a special dynamic "string" data type, on which operations such as concatenation can be carried out.

2.3.2 Useful Notation for Describing Strings

As described above, a string is a sequence of symbols taken from some alphabet. Later, we will need to say such things as:

"suppose x stands for some string taken from the alphabet A".

This is a rather clumsy phrase to have to use. A more accurate, though even clumsier, way of saying it is to say

"x is an element of the set of all strings which can be formed using zero or more symbols of the alphabet A".

There is a convenient and simple notational device to say this. We represent the latter statement as follows:

$$x \in A^*,$$

which relates to the English version as shown in Figure 2.1.

On other occasions, we may wish to say something like:

"x is an element of the set of all strings which can be formed using *one* or more symbols of the alphabet A",

for which we write:

$$x \in A^+$$

which relates to the associated verbal description as shown in Figure 2.2.

Suppose we have the alphabet $\{a, b, c\}$. Then $\{a, b, c\}^*$ is the set

$$\{\varepsilon, a, b, c, aa, ab, ac, ba, bb, bc, ca, cb, cc, aaa, aab, aac, aba, abb, abc, \ldots\}.$$

Clearly, for any non-empty alphabet (i.e. an alphabet consisting of one or more symbols), the set so defined will be *infinite*.

Earlier in the chapter, we discussed the notion of a program printing out the elements of a *finite* set, one by one, terminating when all of the elements

Figure 2.1 How we specify an unknown, possibly empty, string.

Figure 2.2 How we specify an unknown, non-empty, string.

Table 2.2 Systematically printing out all strings in A^*.

```
begin
            <print some symbol to represent the empty string>
            i := 1
            while i >= 0 do
                    <print each of the strings of length i>
                    i := i + 1
            endwhile
end
```

of the set had been printed. If A is some alphabet, we could write a program to print out all the strings in A^*, one by one, such that each string only gets printed out once. Obviously, such a program would *never* terminate (because A^* is an *infinite* set), but we could design the program so that any string in A^* would appear within a finite period of time. Table 2.2 shows a possible method for doing this (as an exercise, you might like to develop the method into a program in your favourite programming language). The method is suggested by the way the first few elements of the set A^*, for $A= \{a, b, c\}$ were written down, above.

An infinite set for which we can print out any given element within a finite time of starting the program is known as a *countably infinite* set. I suggest you think carefully about the program in Table 2.2, as it may help you to appreciate just what is meant by the terms "infinite" and "finite". Clearly, the program specified in Table 2.2 would never terminate. However, on each iteration of the loop, i would have a finite value, and so any string printed out would be finite in length (a necessary condition for a string). Moreover, any string in A^* would appear after a finite period of time.

2.4 Formal Languages

Now we know how to express the notion of all of the strings that can be formed by using symbols from an alphabet, we are in a position to describe what is meant by the term *formal language*. Essentially, a formal language is simply any set of strings formed using the symbols from any alphabet. In set parlance, given some alphabet A,

> a *formal language* is "any (proper or non-proper) subset of the set of all strings which can be formed using zero or more symbols of the alphabet A".

The formal expression of the above statement can be seen in Figure 2.3.

Figure 2.3 The definition of a *formal language.*

A *proper* subset of a set is not allowed to be the *whole* of a given set. For example, the set $\{a, b, c\}$ is a proper subset of the set $\{a, b, c, d\}$, but the set $\{a, b, c, d\}$ is not.

A *non-proper* subset is a subset that is allowed to be the whole of a set. So, the above definition says that, for a given alphabet, A, A^* is a formal language, and so is any subset of A^*. Note that this also means that the empty set, written "$\{\}$" (sometimes written as $\emptyset$) is also a formal language, since it's a subset of A^* (the empty set is a subset of *any* set).

A formal language, then, is any set of strings. To indicate that the strings are part of a language, we usually call them *sentences.* In some books, sentences are called *words.* However, while the strings we have seen so far are similar to English words, in that they are unbroken sequences of alphabetic symbols (e.g. *abca*), later we will see strings that are statements in a programming language, such as

```
if i > 1 then x := x + 1.
```

It seems peculiar to call a statement such as this a "word".

2.5 Methods for Defining Formal Languages

Our definition of a formal language as being a set of strings that are called *sentences* is extremely simple. However, it does not allow us to say anything about the form of sentences in a particular language. For example, in terms of our definition, the Pascal programming language, by which we mean "the set of all syntactically correct Pascal programs", is a subset of the set of all strings which can be formed using symbols found in the character set of a typical computer. This definition, though true, is not particularly helpful if we want to write Pascal programs. It tells us nothing about what makes one string a Pascal program, and another string not a Pascal program, except in the trivial sense that we can immediately rule out any strings containing symbols that are not in the character set of the computer. You would be most displeased if, in attempting to learn to program in Pascal, you opened the Pascal manual to find that it consisted entirely of one statement which said: "Let C be the set of all characters available on the computer. Then the set of compilable Pascal programs, P, is a subset of C^*."

One way of informing you what constitutes "proper" Pascal programs would be to write all the proper ones out for you. However, this would also be unhelpful, albeit in a different way, since such a manual would be infinite, and thus could never be completed. Moreover, it would be a rather tedious process to find the particular program you required.

In this section we discover three approaches to defining a formal language. Following this, every formal language we meet in this book will be defined according to one or more of these approaches.

2.5.1   Set Definitions of Languages

Since a language is a *set* of strings, the obvious way to describe some language is by providing a set definition. Set definitions of the formal languages in which we are interested are of three different types, as now discussed.

The first type of set definition we consider is only used for the smallest finite languages, and consists of writing the language out in its entirety. For example,

$$\{\varepsilon, abc, abbba, abca\}$$

is a language consisting of exactly four strings.

The second method is used for infinite languages, but those in which there is some obvious pattern in all of the strings that we can assume the reader will induce when presented with sufficient instances of that pattern. In this case, we write out sufficient sentences for the pattern to be made clear, then indicate that the pattern should be allowed to continue indefinitely, by using three dots "...". For example,

$$\{ab, aabb, aaabbb, aaaabbbb, \ldots\}$$

suggests the infinite language consisting of all strings which consist of one or more as followed by one or more bs and in which the number of as equals the number of bs.

The final method, used for many *finite* and *infinite* languages, is to use a set definition to specify how to construct the sentences in the language, i.e., provide a function to deliver the sentences as its output. In addition to the function itself, we must provide a specification of how many strings should be constructed. Such set definitions have the format shown in Figure 2.4.

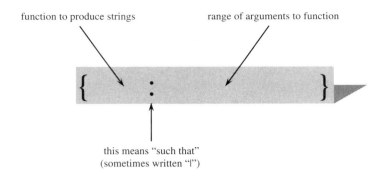

function to produce strings   range of arguments to function

this means "such that"
(sometimes written "|")

Figure 2.4 Understanding a set definition of a formal language.

For the "function to produce strings", of Figure 2.4, we use combinations of the string functions we considered earlier (*index*, *power* and *concatenation*). A language that was defined immediately above,

"all strings which consist of one or more as followed by one or more bs and in which the number of as equals the number of bs"

can be defined using our latest method as:

$$\{a^i b^i : i \geq 1\}.$$

The above definition is explained in Table 2.3.

From Table 2.3 we can see that $\{a^i b^i : i \geq 1\}$ means:

"the set of all strings consisting of i copies of a followed by i copies of b such that i is allowed to take on the value of each and every whole number value greater than or equal to 1".

Table 2.3 What the set definition $\{a^i b^i : i \geq 1\}$ means.

Notation	String function	Meaning
a	N/A	"the string a"
b	N/A	"the string b"
a^i	power	"the string formed by writing down i copies of the string a"
b^i	power	"the string formed by writing down i copies of the string b"
$a^i b^i$	concatenation	"the string formed by writing down i copies of a followed by i copies of b"
$: i \geq 1$	N/A	"such that i is allowed to take on the value of each and every whole number value greater than or equal to 1 (we could have written $i > 0$)"

Changing the right-hand side of the set definition can change the language defined. For example $\{a^i b^i : i \geq 0\}$ defines:

> "the set of all strings consisting of i copies of a followed by i copies of b such that i is allowed to take on the value of each and every whole number value greater than or equal to 0".

This latter set is our original set, along with the empty string (since $a^0 = \varepsilon$, $b^0 = \varepsilon$, and therefore $a^0 b^0 = \varepsilon\varepsilon = \varepsilon$). In set parlance, $\{a^i b^i : i \geq 0\}$ is the *union* of the set $\{a^i b^i : i \geq 1\}$ with the set $\{\varepsilon\}$, which can be written:

$$\{a^i b^i : i \geq 0\} = \{a^i b^i : i \geq 1\} \cup \{\varepsilon\}.$$

The immediately preceding example illustrates a further useful feature of sets. We can often simplify the definition of a language by creating several sets and using the union, intersection and set difference operators to combine them into one. This sometimes removes the need for a complicated expression in the right-hand side of our set definition. For example, the definition

$$\{a^i b^j c^k : i \geq 1, j \geq 0, k \geq 0, \ if \ i \geq 3 \ then \ j = 0 \ else \ k = 0\},$$

is probably better represented as

$$\{a^i c^j : i \geq 3, j \geq 0\} \cup \{a^i b^j : 1 \leq i < 3, j \geq 0\},$$

which means

> "the set of strings consisting of 3 or more *a*s followed by zero or more *c*s, or consisting of 1 or 2 *a*s followed by zero or more *b*s".

2.5.2   Decision Programs for Languages

We have seen how to define a language by using a formal set definition. Another way of describing a language is to provide a program that tells us whether or not any given string of symbols is one of its sentences. Such a program is called a *decision program*. If the program always tells us, for any string, whether or not the string is a sentence, then the program in an implicit sense defines the language, in that the language is the set containing each and every string that the program tells us is a sentence. That is why we use a special term, "sentence", to describe *a string that belongs to a language*. A *string* input to the program may or may not be a *sentence* of the language; the program should tell us. For an alphabet A, a language is any *subset* of A^*. For any interesting language, then, there will be many strings in A^* that are not sentences.

Later in this book we will be more precise about the form these decision programs take, and what can actually be achieved with them. For now, however, we will consider an example to show the basic idea.

If you have done any programming at all, you will have used a decision program on numerous occasions. The decision program you have used is a component of the *compiler*. If you write programs in a language such as Pascal, you submit your program text to a compiler, and the compiler tells you if the text is a syntactically correct Pascal program. Of course, the compiler does a lot more than this, but a very important part of its job is to tell us if the source text (string) is a syntactically correct Pascal program, i.e. a *sentence* of the *language* called "Pascal".

Consider again the language

$$\{a^i c^j : i \geq 3, j \geq 0\} \cup \{a^i b^j : 1 \leq i < 3, j \geq 0\},$$

i.e.,

"the set of strings consisting of 3 or more *a*s followed by zero or more *c*s, or consisting of 1 or 2 *a*s followed by zero or more *b*s".

Table 2.4 shows a decision program for the language.

The program of Table 2.4 is purely for illustration. In the next chapter we consider formal languages for which the above type of decision program can be created automatically. For now, examine the program to convince yourself that it correctly meets its specification, which can be stated as follows:

"given any string in $\{a, b, c\}^*$, tell us whether or not that string is a sentence of the language

$$\{a^i c^j : i \geq 3, j \geq 0\} \cup \{a^i b^j : 1 \leq i < 3, j \geq 0\}".$$

2.5.3   Rules for Generating Languages

We have seen how to describe formal languages by providing set definitions and we have encountered the notion of a decision program for a language. The third method, which is the basis for the remainder of this chapter, defines a language by providing a set of rules to generate sentences of a language. We require that such rules are able to generate every one of the sentences of a language, and no others. Analogously, a set definition describes every one of the sentences, and no others, and a decision program says "yes" to every one of the sentences, and to no others.

Table 2.4 A decision program for a formal language.

1:	read(sym)		{assume <u>read</u> just gives us the next symbol in the string being examined}
	case sym of		
	eos:	goto N	{assume read returns special symbol "eos" if at end of string}
	"a":	goto 2	
	"b":	goto N	
	"c":	goto N	
	endcase		{case statement selects between alternatives as in Pascal}
2:	read(sym)		
	case sym of		
	eos:	goto Y	{if we get here we have a string of one a which is OK}
	"a":	goto 3	
	"b":	goto 6	{we can have a b after one a}
	"c":	goto N	{any cs must follow three or more as – here we've only had one}
	endcase		
3:	read(sym)		
	case sym of		
	eos:	goto Y	{if we get here we've read a string of two as which is OK}
	"a":	goto 4	
	"b":	goto 6	{we can have a b after two as}
	"c":	goto N	{any cs must follow three or more as – here we've only had two}
	endcase		
4:	read(sym)		
	case sym of		
	eos:	goto Y	{if we get here we've read a string of three or more as which is OK}
	"a":	goto 4	{we loop here because we allow any number of as ≥ 3}
	"b":	goto N	{b can only follow one or two as}
	"c":	goto 5	{cs are OK after three or more as}
	endcase		
5:	read(sym)		
	case sym of		
	eos:	goto Y	{if we get here we've read ≥ 3 as followed by ≥ 1 cs which is OK}
	"a":	goto N	{as after cs are not allowed}
	"b":	goto N	{bs are only allowed after one or two as}
	"c":	goto 5	{we loop here because we allow any number of cs after ≥ 3 as}
	endcase		
6:	read(sym)		
	case sym of		
	eos:	goto Y	{we get here if we've read 1 or 2 as followed by ≥ 1 bs – OK}
	"a":	goto N	{no as allowed after bs}

Table 2.4 (continued)

	"b":	goto 6	{we loop here because we allow any number of *b*s after 1 or 2 *a*s}
	"c":	goto N	{no *c*s are allowed after *b*s}
	endcase		
Y:	write("yes")		
	goto E		
N:	write("no")		
	goto E		
E:	{end of program}		

There are several ways of specifying rules to generate sentences of a language. One popular form is the *syntax diagram*. Such diagrams are often used to show the structure of programming languages, and thus inform you how to write syntactically correct programs (*syntax* is considered in more detail in Chapter 3).

Figure 2.5 shows a syntax diagram for the top level syntax of the Pascal "program" construct.

The diagram in Figure 2.5 tells us that the syntactic element called a "program" consists of

the string "PROGRAM" (entities in rounded boxes and circles represent actual strings that are required at a given point),

followed by something called

an "identifier" (entities in rectangles are those which need elaborating in some way that is specified in a further definition),

followed by

an open bracket "(",

followed by

a list of one or more "identifiers", in which *every one except the last* is followed by a comma, ",", followed by a semi-colon, ";",

followed by

a close bracket, ")",

program

Figure 2.5 Syntax diagram for the Pascal construct "program".

followed by

something called a "block",

followed by

a full stop, ".".

In Figure 2.6 we see the syntax diagram for the entity "identifier".

Figure 2.6 shows us that an "identifier" consists of a letter followed by zero or more letters and/or digits.

The following fragment of Pascal:

```
program calc(input, output, infile26, outfile23);
```

associates with the syntax diagram for "program" as shown in Figure 2.7.

Of course, the diagrams in Figures 2.5 and 2.6, together with all of the other diagrams defining the syntax of Pascal, cannot tell us how to write a program to solve a given problem. That is a *semantic* consideration, relating to the *meaning* of the program text, not only its *form*. The diagrams merely describe the *syntactic structure* of constructs belonging to the Pascal language.

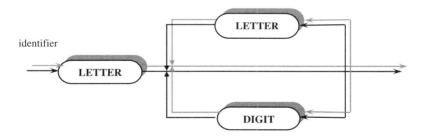

Figure 2.6 Syntax diagram for a Pascal "identifier".

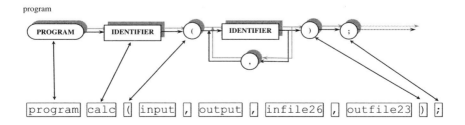

Figure 2.7 How a syntax diagram describes a Pascal statement.

Table 2.5 BNF version of Figures 2.5 and 2.6.

<program> ::= <program heading> <block>.
<program heading> :=<u>program</u><identifier> (<identifier> { , <identifier> }) ;
<identifier> ::= <letter> {<letter or digit>}
<letter or digit> ::= <letter> | <digit>

An alternative method of specifying the syntax of a programming language is to use a notation called Backus-Naur form (BNF).[1] Table 2.5 presents a BNF version of our syntax diagrams from above.

The meaning of the notation in Table 2.5 should be reasonably clear when you see its correspondence with syntax diagrams, as shown in Figure 2.8.

Formalisms such as syntax diagrams and BNF are excellent ways of defining the syntax of a language. If you were taught to use a programming language, you may never have looked at a formal definition of its syntax. Analogously, you probably did not learn your own "natural" language by studying a book describing its grammar. However, many programming languages are similar to each other in many respects, and learning a subsequent programming language is made easier if the syntax is clearly defined. Syntax descriptions can also be useful for refreshing your memory about the syntax of a programming language with which you are familiar, particularly for types of statements you rarely use.

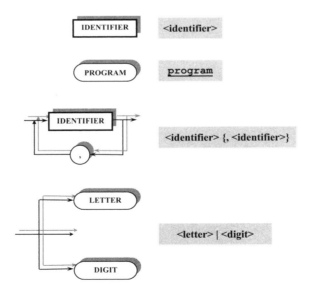

Figure 2.8 How syntax diagrams and BNF correspond.

---

[1] The formalism we describe here is actually *Extended* BNF (EBNF). The original BNF did not include the repetition construct found in Table2.5.

If you want to see how concisely a whole programming language can be described in BNF, see the original definition of the Pascal language,[2] from where the above Pascal syntax diagrams and BNF descriptions were obtained. The BNF definitions for the whole Pascal language are presented in only *five* pages.

2.6 Formal Grammars

A *grammar* is a set of rules for generating strings. The grammars we will use in the remainder of this book are known as *phrase structure grammars* (*PSG*s). Here, our formal definitions will be illustrated by reference to the following grammar:

$$S \to aS \mid bB$$

$$B \to bB \mid bC \mid cC$$

$$C \to cC \mid c.$$

In order to use our grammar, we need to know something about the status of the symbols that we have used. Table 2.6 provides an informal description of the symbols that appear in grammars such as the one above.

Table 2.6 The symbols that make up the Phrase Structure Grammar:

$S \to aS \mid bB$

$B \to bB \mid bC \mid cC$

$C \to cC \mid c.$

Symbols	Name and meaning
S, B, C	*non-terminal* symbols
	[BNF: things in angled brackets e.g. <identifier>]
S	special non-terminal, called a *start*, or *sentence*, symbol
	[BNF: in our example above, <program>]
a, b, c	*terminal* symbols: only these symbols can appear in sentences
	[BNF: the underlined terms (e.g. program) and punctuation symbols (e.g. ";")]
$\to$	*production arrow*
	[BNF: the symbol "::="]
$S \to aS$	*production rule*, usually called simply a *production* (or sometimes we'll just use the word *rule*). Means "*S produces aS*", or "*S can be replaced by aS*". The string to the left of $\to$ is called the *left-hand side* of the production, the string to the right of $\to$ is called the *right-hand side*.
	[BNF: this rule would be written as <S> ::= a<S>]
|	"or", so $B \to bB \mid bC \mid cC$ means "*B produces bB or bC or cC*". Note that this means that $B \to bB \mid bC \mid cC$ is really *three* production rules i.e. $B \to bB$, $B \to bC$, and $B \to cC$. So there are *seven* production rules altogether in the example grammar above.
	[BNF: exactly the same]

---

[2] Jensen and Wirth (1975) – see Further Reading section.

2.6.1 Grammars, Derivations and Languages

Table 2.7 presents an informal description, supported by examples using our grammar above, of how we use a grammar to generate a sentence.

As you can see from Table 2.7, there is often a choice as to which rule to apply at a given stage. For example, when the resulting string was aaS, we could have applied the rule $S \to aS$ as many times as we wished (adding another a each time). A similar observation can be made for the applicability of the $C \to cC$ rule when the resulting string was $aabcC$, for example.

Here are some other strings we could create, by applying the rules in various ways:

$$abcc,$$

$$bbbbc, \text{ and}$$

$$a^3 b^2 c^5.$$

You may like to see if you can apply the rules yourself to create the above strings. You must always begin with a rule that has S on its left-hand side (that is why S is called the *start* symbol).

We write down the S symbol to start the process, and we merely repeat the process described in Table 2.7 as

if a substring of the resulting string matches the left-hand side of one or more productions, replace that substring by the right-hand side of any one of those productions,

until the following becomes true

if the resulting string consists entirely of terminals,

Table 2.7 Using a Phrase Structure Grammar.

Action taken	Resulting string	Production applied
Start with S, the *start* symbol	S	
If a substring of the resulting string matches the left-hand side of one or more productions, replace that substring by the right-hand side of any one of those productions	aS	$S \to aS$
,,	aaS	$S \to aS$
,,	$aabB$	$S \to bB$
,,	$aabcC$	$B \to cC$
,,	$aabccC$	$C \to cC$
,,	$aabccc$	$C \to c$
If the resulting string consists entirely of terminals, then stop.		

at which point we:

stop.

You may wonder why the process of matching the substring was not presented as:

if a non-terminal symbol *in the resulting string matches the left-hand side of one or more productions, replace that* non-terminal symbol *by the right-hand side of any one of those productions.*

This would clearly work for the example grammar given. However, as discussed in the next section, grammars are not necessarily restricted to having single non-terminals on the left-hand sides of their productions.

The process of creating strings using a grammar is called *deriving* them, so when we show how we've used the grammar to *derive* a string (as was done in Table 2.7), we're showing a *derivation* for (or *of*) that string.

Let us now consider all of the "terminal strings" – strings consisting entirely of terminal symbols, also known as *sentences*– that we can use the example grammar to derive. As this is a simple grammar, it's not too difficult to work out what they are.

Figure 2.9 shows the choice of rules possible for deriving terminal strings from the example grammar.

Any "legal" application of our production rules, starting with S, the start symbol, alone, and resulting in a *terminal string*, would involve us in following a path through the diagram in Figure 2.9, starting in Box 1, passing through Box 2, and ending up in Box 3. The boxes in Figure 2.9 are annotated with the strings produced by taking given options in applying the rules. Table 2.8 summarises the strings described in Figure 2.9.

We now define a set that contains all of the terminal strings (and only those strings) that can be derived from the example grammar. The set will contain all strings defined as follows:

A string taken from the set $\{a^i b: i \geq 0\}$ concatenated with a string taken from the set $\{b^j \; : \; j \geq 1\} \cup \{b^j c \; : \; j \geq 0\}$ concatenated with a string taken from the set $\{c^k: k \geq 1\}$.

The above can be written as:

$$\{a^i bb^j c^k : i \geq 0, j \geq 1, k \geq 1\} \cup \{a^i bb^j cc^k : i \geq 0, j \geq 0, k \geq 1\}.$$

Observe that $bb^j, j \geq 1$ is the same as $b^j, j \geq 2$, and $bb^j, j \geq 0$ is the same as $b^j, j \geq 1$, and $cc^k, k \geq 1$ is the same as $c^k, k \geq 2$ so we could write:

$$\{a^i b^j c^k : i \geq 0, j \geq 2, k \geq 1\} \cup \{a^i b^j c^k : i \geq 0, j \geq 1, k \geq 2\}.$$

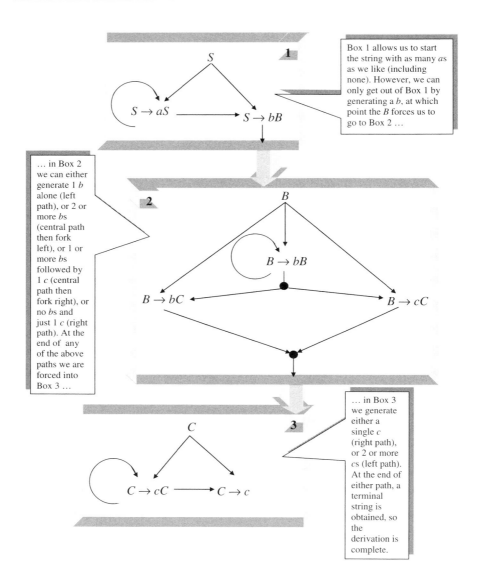

Figure 2.9 Working out all of the terminal strings that a grammar can generate.

This looks rather complicated, but essentially there is only one awkward case, which is that if there is only one b then there must be 2 or more cs (any more than 1 b and we can have 1 or more cs). So we could have written:

$$\{a^i b^j c^k : i \geq 0, j \geq 1, k \geq 1,\ if\ j = 1\ then\ k \geq 2\ else\ k\ \geq 1\}.$$

Table 2.8 The language generated by a grammar.

Box in Figure 2.9.	Informal description of derived strings	Formal description of derived strings
$\boxed{1}$ i.e. productions $S \rightarrow aS \mid S \rightarrow bB$	"any non-zero number of as followed by bB" or "just bB " which is the same as saying "zero or more as followed by bB "	$a^i bB,\ i \geq 0$ ...the B at the end...
...is expanded in Box 2...		
$\boxed{2}$ *i.e. productions* $B \rightarrow bB \mid bC \mid cC$	"any non-zero number of bs followed by either bC or cC" or "just bC" or "just cC"	$b^j C,\ j \geq 1$ or $b^j cC,$ $j \geq 0$ ...the C at the end...
...is expanded in Box 3...		
$\boxed{3}$ i.e. productions $C \rightarrow cC \mid c$	"any non-zero number of cs followed by one c" or "just one c" which is the same as saying "one or more cs"	$c^k,\ k \geq 1$

Whichever way we write the set, one point should be made clear: the set is a set of strings formed from symbols in the alphabet $\{a, b, c\}^*$, that is to say, the set is a *formal language*.

2.6.2 The Relationship between Grammars and Languages

We are now ready to give an intuitive definition of the relationship between grammars and languages:

> The *language generated by a grammar* is the set of all *terminal strings* that can be *derived* using the *productions* of that grammar, each derivation beginning with the *start symbol* of that grammar.

Our example grammar, when written like this:

$$S \rightarrow aS \mid bB$$
$$B \rightarrow bB \mid bC \mid cC$$
$$C \rightarrow cC \mid c$$

is not fully defined. A grammar is fully defined when we know which symbols are terminals, which are non-terminals, and which of the non-terminals is the start symbol. In this book, we will usually see only the productions of a grammar, and we will assume the following:

- capitalised letters are *non-terminal symbols*

- non-capitalised letters are *terminal symbols*

- the capital letter *S* is the *start symbol*.

The above will always be the case unless explicitly stated otherwise.

2.7 Phrase Structure Grammars and the Chomsky Hierarchy

The production rules of the example grammar from the preceding section are simple in format. For example, the left-hand sides of all the productions consist of lone non-terminals. As we see later in the book, restricting the form of productions allowed in a grammar in certain ways simplifies certain language processing tasks, but it also reduces the sophistication of the languages that such grammars can generate. For now, we will define a scheme for classifying grammars according to the "shape" of their productions which will form the basis of our subsequent discussion of grammars and languages. The classification scheme is called the Chomsky hierarchy, named after Noam Chomsky, an influential American linguist.

2.7.1 Formal Definition of Phrase Structure Grammars

To prepare for specifying the Chomsky hierarchy, we first need to precisely define the term *phrase structure grammar* (PSG). Table 2.9 does this.

Formally, then, a PSG, *G*, is specified as (N, T, P, S). This is what mathematicians call a "tuple" (of four elements).

The definition in Table 2.9 makes it clear that the empty string, ε, cannot appear alone on the left-hand side of any of the productions of a PSG. Moreover, the definition tells us that ε *is* allowed on the right-hand side. Otherwise, any strings of terminals and/or non-terminals can appear on either side of productions. However, in most grammars we usually find that there are one or more non-terminals on the *left-hand side* of each production.

As we always start a derivation with a lone *S* (the *start* symbol), for a grammar to derive anything it must have at least one production with S alone on its left-hand side. This last piece of information is not specified in the definition above, as there is nothing in the formal definition of PSGs that says they *must* generate

Table 2.9 The formal definition of a phrase structure grammar.

Any PSG, G, consists of the following:		
N	a set of *non-terminal* symbols	an alphabet, containing no symbols that can appear in sentences
T	a set of *terminal* symbols	also an alphabet, containing only symbols that *can* appear in sentences
P	a set of *production rules* of the form	this specification uses the notation for specifying strings from an alphabet we looked at earlier.
	$x \to y$, where $x \in (N \cup T)^+$, and $y \in (N \cup T)^*$	x is the left-hand side of a production, y the right-hand side. The definition of y means: *the right-hand side of each production is a possibly empty string of terminals and/or non-terminals.* The only difference between the specification above and the one for x (the left-hand side) is that the one for x uses "+" rather than "*". So the specification for x means: *the left-hand side of each production is a non-empty string of terminals and/or non-terminals.*
S	a member of N, designated as the *start*, or *sentence* symbol	the non-terminal symbol with which we always begin a derivation

anything. To refer back to our earlier example grammar, its full formal description would be as shown in Table 2.10.

2.7.2 Derivations, Sentential Forms, Sentences and "L(G)"

We have formalised the definition of a *phrase structure grammar* (PSG). We now formalise our notion of derivation, and introduce some useful terminology to support subsequent discussion. To do this, we consider a new grammar:

$$S \to aB \mid bA \mid \varepsilon$$
$$A \to aS \mid bAA$$
$$B \to bS \mid aBB.$$

Table 2.10 The (N, T, P, S) form of a grammar.

Productions	(N, T, P, S)	
	($\{S, B, C\}$,	- - - N
$S \to aS \mid bB$	$\{a, b, c\}$,	- - - T
$B \to bB \mid bC \mid cC$	$\{S \to aS, \ S \to bB, B \to bB, B \to bC,$	- - - P
$C \to cC \mid c$	$B \to cC, C \to cC, C \to c\}$,	
	S	- - - S
	)	

Using the conventions outlined earlier, we know that S is the start symbol, $\{S, A, B\}$ is the set of non-terminals (N), and $\{a, b\}$ is the set of terminals (T). So we need not provide the full (N, T, P, S) definition of the grammar.

As in our earlier example, the left-hand sides of the above productions all consist of single non-terminals. We see an example grammar that differs from this later in the chapter.

Here is a string in $(N \cup T)^+$ that the above productions can be used to derive, as you might like to verify for yourself:

$$abbbaSA.$$

This is not a terminal string, since it contains non-terminals (S and A). Therefore it is not a sentence. The next step could be, say, to apply the production $A \to bAA$, which would give us

$$abbbaSbAA,$$

which is also not a sentence.

We now have two strings, $abbbaSA$ and $abbbaSbAA$ that are such that the *former can be used as a basis for the derivation of the latter by the application of one production rule of the grammar*. This is rather a mouthful, even if we replace *"by the application of one production rule of the grammar"* by the phrase *"in one step"*, so we introduce a symbol to represent this relationship. We write:

$$abbbaSA \Rightarrow abbbaSbAA.$$

To be absolutely correct, we should give our grammar a name, say G, and write

$$abbbaSA \Rightarrow^G abbbaSbAA$$

to denote which particular grammar is being used. Since it is usually clear in our examples which grammar is being used, we will simply use $\Longrightarrow$. We now use this symbol to show how our example grammar derives the string $abbbaSbAA$:

$$S \Rightarrow aB$$
$$aB \Rightarrow abS$$
$$abS \Rightarrow abbA$$
$$abbA \Rightarrow abbbAA$$
$$abbbAA \Rightarrow abbbaSA$$
$$abbbaSA \Rightarrow abbbaSbAA$$

As it is tedious to write out each intermediate stage twice, apart from the first (S) and the last ($abbbaSbAA$), we allow an abbreviated form of such a derivation as follows:

$$S \Rightarrow aB \Rightarrow abS \Rightarrow abbA \Rightarrow abbbAA \Rightarrow abbbaSA \Rightarrow abbbaSbAA.$$

We now use our new symbol as the basis of some additional useful notation, as shown in Table 2.11.

A new term is now introduced to simplify references to the intermediate stages in a derivation. We call these intermediate stages *sentential forms*. Formally, given any grammar, G, a *sentential form* is any string that can be derived in zero or more steps from the start symbol, S. By "any string", we mean exactly that; not only terminal strings, but any string of terminals and/or non-terminals. Thus, *a sentence is a sentential form*, but *a sentential form is not necessarily a sentence*. Given the simple grammar

$$S \to aS | a,$$

some sentential forms are: S, $aaaaaaS$ and a^{10}. Only one of these sentential forms (a^{10}) is a sentence, as it's the only one that consists entirely of *terminal symbols*.

Formally, using our new notation,

Table 2.11 Useful notation for discussing derivations, and some example true statements for the grammar:

$$S \to aB \mid bA \mid \varepsilon$$
$$A \to aS \mid bAA$$
$$B \to bS \mid aBB.$$

Notation	Meaning	Example true statements
$x \Rightarrow y$	the application of one production rule results in the string x becoming the string y also expressed as "*x generates y* in one step", or "*x produces y* in one step", or "*y is derived from x* in one step"	$aB \Rightarrow abS$ $S \Rightarrow \varepsilon$ $abbbaSA \Rightarrow abbba\ SbAA$
$x \Rightarrow^* y$	x generates y in zero or more steps, or just "*x generates y*", or "*x produces y*", or "*y is derived from x*"	$S \Rightarrow^* S$ $S \Rightarrow^* abbbaSA$ $aB \Rightarrow^* abbbaa$
$x \Rightarrow^+ y$	x generates y in one or more steps, or just "*x generates y*", or "*x produces y*", or "*y is derived from x*"	$S \Rightarrow^+ abbbaSA$ $abbba\ SbAA \Rightarrow^+ abbbabaa$

if $S \Longrightarrow^* x$, then x is a sentential form.

if $S \Longrightarrow^* x$, and x is a terminal string, then x is a *sentence*.

We now formalise a definition given earlier, this being the statement that

the language generated by a grammar is the set of all terminal strings that can be derived using the productions of that grammar, each derivation beginning with the start symbol of that grammar.

Using various aspects of the notation introduced in this chapter, this becomes:

$$\text{Given a PSG, } G, L(G) = \{x : x \in T^* \text{ and } S \Rightarrow^* x\}.$$

(Note that the definition assumes that we have specified the set of terminals and the start symbol of the grammar, which as we said earlier is done implicitly in our examples.)

So, if G is some PSG, $L(G)$ means *the language generated by G*. As the set definition of $L(G)$ clearly states, the set $L(G)$ contains all of the terminal strings generated by G, but only the strings that G generates. It is very important to realise that this is what it means when we say *the language generated by the grammar*.

We now consider three examples, to reinforce these notions. The first is an example grammar encountered above, now labelled G_1:

$$S \rightarrow aS \,|\, bB$$
$$B \rightarrow bB \,|\, bC \,|\, cC$$
$$C \rightarrow cC \,|\, c.$$

We have already provided a set definition of $L(G_1)$; it was:

$$L(G_1) = \{a^i b^j c^k : i \geq 0, j \geq 1, k \geq 1, \text{ if } j = 1 \text{ then } k \geq 2 \text{ else } k \geq 1\}.$$

Another grammar we have already encountered, which we now call G_2, is:

$$S \rightarrow aBj \,|\, bA \,|\, \varepsilon$$
$$A \rightarrow aS \,|\, bAA$$
$$B \rightarrow bS \,|\, aBB.$$

This is more complex than G_1, in the sense that some of G_2's productions have more than one non-terminal on their right-hand sides.

$$L(G_2) = \{x : x \in \{a, b\}^* \text{ and the number of } as \text{ in } x \text{ equals the number of } bs\}.$$

I leave it to you to establish that the above statement is true.

Note that L(G_2) is not the same as a set that we came across earlier, i.e.

$$\{a^i b^i : i \geq 1\},$$

which we will call set A. In fact, set A is a proper subset of $L(G_2)$. G_2 can generate all of the strings in A, but it generates many more besides (such as ε, *bbabb-baaaaab*, and so on). A grammar, G_3, such that $L(G_3) = A$ is:

$$S \rightarrow ab \,|\, aSb.$$

2.7.3 The Chomsky Hierarchy

This section describes a classification scheme for PSGs, and the corresponding phrase structure languages (PSLs) that they generate, which is of the utmost importance in determining certain of their computational features. PSGs can be classified in a hierarchy, the location of a PSG in that hierarchy being an indicator of certain characteristics required by a decision program for the corresponding language. We saw above how one example language could be processed by an extremely simple decision program. Much of this book is devoted to investigating the computational nature of formal languages. We use as the basis of our investigation the classification scheme for PSGs and PSLs called the *Chomsky hierarchy*.

Classifying a grammar according to the Chomsky hierarchy is based solely on the presence of certain patterns in the productions. Table 2.12 shows how to make the classification. The types of grammar in the Chomsky hierarchy are named types 0 to 3, with 0 as the most general type. Each type from 1 to 3 is defined according to one or more restrictions on the definition of the type numerically preceding it, which is why the scheme qualifies as a *hierarchy*.

If you are observant, you may have noticed an anomaly in Table 2.12. Context sensitive grammars are not allowed to have the empty string on the right-hand side of productions, whereas all of the other types are. This means that, for example, our grammar G_2, which can be classified as unrestricted and as context free (but not as regular), cannot be classified as context sensitive. However, every grammar that can be classified as regular can be classified as context free, and every grammar that can be classified as context free can be classified as unrestricted.

Table 2.12 The Chomsky hierarchy.

Type №	Type name	Patterns to which ALL productions must conform	Informal description and examples
0	unrestricted	$x \to y, x \in (N \cup T)^{+}$, $y \in (N \cup T)^{*}$	The definition of PSGs we have already seen. Anything allowed on the left-hand side (except for ε), anything allowed on the right. All of our example grammars considered so far conform to this. Example type 0 production: $aXYpq \to aZpq$ (all productions of G_1, G_2 and G_3 conform – but see below).
1	context sensitive	$x \to y, x \in (N \cup T)^{+}$, $y \in (N \cup T)^{+}$, $\lvert x \rvert \leq \lvert y \rvert$	As for *type* 0, but we are not allowed to have ε on the left- *or the right-hand sides.* Note that the example production given for *type* 0 is <u>not</u> a context sensitive production, as the length of the right-hand side is less than the length of the left. Example type 1 production: $aXYpq \to aZwpq$ (all productions of G_1 and G_3 conform, but not all of those of G_2 do).
2	context free	$x \to y, x \in N$, $y \in (N \cup T)^{*}$	Single non-terminal on left, any mixture of terminals and/or non-terminals on the right. Also, ε is allowed on the right. Example type 2 production: $X \to XapZQ$ (all productions of G_1, G_2, *and* G_3 conform).
3	regular	$w \to x$, *or* $w \to yz$ $w \in N$, $x \in T \cup \{\varepsilon\}$, $y \in T$, $z \in N$	Single non-terminal on left, and either ■ ε or a single terminal, or ■ a single terminal followed by a single non-terminal, on the right. Example type 3 productions: $P \to pQ,$ $F \to a$ all of the productions of G_1 conform to this, but G_2 and G_3 do not.

When classifying a grammar according to the Chomsky hierarchy, you should remember the following:

For a grammar to be classified as being of a certain type, *each and every production of that grammar must match the pattern specified for productions of that type.*

Which means that the following grammar:

$$S \rightarrow aS \mid aA \mid AA$$
$$A \rightarrow aA \mid a,$$

is classified as *context free*, since the production $S \rightarrow AA$ does not conform to the pattern for regular productions, even though all of the other productions do.

So, given the above rule that all productions must conform to the pattern, you classify a grammar, G, according to the procedure in Table 2.13.

Table 2.13 tells us to begin by attempting to classify G according to the most restricted type in the hierarchy. This means that, as indicated by Table 2.12, G_1 is a *regular* grammar, and G_2 and G_3 are *context free* grammars. Of course, we know that as *all* regular grammars are context free grammars, G_1 is also context free. Similarly, we know that they can *all* be classified as unrestricted. But we make the classification as specific as possible.

From the above, it can be seen that classifying a PSG is done simply by seeing if its productions match a given pattern. As we already know, grammars generate languages. In terms of the Chomsky hierarchy, *a language is of a given type if it is generated by a grammar of that type.* So, for example,

$$\{a^i b^i : i \geq 1\} \qquad \text{(set } A \text{ mentioned above)}$$

is a context free language, since it is generated by G_3, which is classified as a context free grammar. However, how can we be sure that there is not a *regular grammar* that could generate A? We see later on that the more restricted the language (in the Chomsky hierarchy), the simpler the decision program for the language. It is therefore useful to be able to define the simplest possible type of grammar

Table 2.13 The order in which to attempt the classification of a grammar, G, in the Chomsky hierarchy.

```
if G is regular then
      return("regular")
else
      if G is context free then
            return("context free")
      else
            if G is context sensitive then
                  return("context sensitive")
            else
                  return("unrestricted")
            endif
      endif
endif
```

for a given language. In the meantime, you might like to see if you can create a *regular* grammar to generate set A (clue: do not devote too much time to this!).

From a theoretical perspective, the immediately preceding discussion is very important. If we can establish that there are languages that can be generated by grammars at some level of the hierarchy and cannot be generated by more restricted grammars, then we are sure that we do indeed have a genuine *hierarchy*. However, there are also *practical* issues at stake, for as mentioned above, and discussed in more detail in Chapters 4, 5 and 7, each type of grammar has associated with it a type of decision program, in the form of an abstract machine. The more restricted a language is, the simpler the type of decision program we need to write for that language.

In terms of the Chomsky hierarchy, our main interest is in *context free* languages, as it turns out that the syntactic structure of most programming languages is represented by context free grammars. The grammars and languages we have looked at so far in this book have all been context free (remember that any *regular* grammar or language is, by definition, also context free).

2.8 A Type 0 Grammar: Computation as Symbol Manipulation

We close this chapter by considering a grammar that is more complex than our previous examples. The grammar, which we label G_4, has productions as follows (each row of productions has been numbered, to help us to refer to them later).

$$S \to AS \,|\, AB \qquad\qquad (1)$$
$$B \to BB \,|\, C \qquad\qquad (2)$$
$$AB \to HXNB \qquad\qquad (3)$$
$$NB \to BN \qquad\qquad (4)$$
$$BM \to MB \qquad\qquad (5)$$
$$NC \to Mc \qquad\qquad (6)$$
$$Nc \to Mcc \qquad\qquad (7)$$
$$XMBB \to BXNB \qquad\qquad (8)$$
$$XBMc \to Bc \qquad\qquad (9)$$
$$AH \to HA \qquad\qquad (10)$$
$$H \to a \qquad\qquad (11)$$
$$B \to b \qquad\qquad (12)$$

G_4 is a type 0, or unrestricted grammar. It would be context sensitive, but for the production $XBMc \rightarrow Bc$, which is the only production with a right-hand side shorter than its left-hand side.

Table 2.14 represents the derivation of a particular sentence using this grammar. It is presented step by step. Each sentential form, apart from the *sentence* itself, is followed by the number of the row in G_4 from which the production used to achieve the next step was taken. Table 2.14 should be read row by row, left to right.

The sentence derived is $a^2 b^3 c^6$. Notice how, in Table 2.14, the grammar replaces each A in the sentential form $AABBBC$ by H, and each time it does this it places one c at the rightmost end for each B. Note also how the grammar uses non-terminals as "markers" of various types:

- H is used to replace the As that have been accounted for

- X is used to indicate how far along the Bs we have reached

- N is used to move right along the Bs, each time ending in a c being added to the end of the sentential form

- M is used to move left back along the Bs.

You may also notice that at many points in the derivation several productions are applicable. However, many of these productions lead eventually to "dead ends", i.e., sentential forms that cannot lead eventually to sentences.

Table 2.14 A type 0 grammar is used to derive a sentence.

STAGE	row	STAGE	row	STAGE	row
S	(1)	AS	(1)	AAB	(2)
$AABB$	(2)	$AABBB$	(2)	$AABBBC$	(3)
$AHXNBBBC$	(4)	$AHXBNBBC$	(4)	$AHXBBNBC$	(4)
$AHXBBBNC$	(6)	$AHXBBBMc$	(5)	$AHXBBMBc$	(5)
$AHXBMBBc$	(5)	$AHXMBBBc$	(8)	$AHBXNBBc$	(4)
$AHBXBNBc$	(4)	$AHBXBBNc$	(7)	$AHBXBBMcc$	(5)
$AHBXBMBcc$	(5)	$AHBXMBBcc$	(8)	$AHBBXNBcc$	(4)
$AHBBXBNcc$	(7)	$AHBBXBMccc$	(9)	$AHBBBccc$	(10)
$HABBBccc$	(3)	$HHXNBBBccc$	(4)	$HHXBNBBccc$	(4)
$HHXBBNBccc$	(4)	$HHXBBBNccc$	(7)	$HHXBBBMcccc$	(5)
$HHXBBMBcccc$	(5)	$HHXBMBBcccc$	(5)	$HHXMBBBcccc$	(8)
$HHBXNBBcccc$	(4)	$HHBXBNBcccc$	(4)	$HHBXBBNcccc$	(7)
$HHBXBBMccccc$	(5)	$HHBXBMBccccc$	(5)	$HHBXMBBccccc$	(8)
$HHBBXNBccccc$	(4)	$HHBBXBNccccc$	(7)	$HHBBXBMcccccc$	(9)
$HHBBBcccccc$	(11)	$aHBBBcccccc$	(11)	$aaBBBccccccc$	(12)
$aabBBccccccc$	(12)	$aabbBcccccc$	(12)	$aabbbcccccc$	

The language generated by G_4, i.e. $L(G_4)$, is $\{a^i b^j c^{i \times j} : i, j \geq 1\}$. This is the set:

"all strings of the form one or more as followed by one or more bs followed by cs in which the number of cs is the number of as *multiplied* by the number of bs".

You may wish to convince yourself that this is the case.

G_4 is rather a complicated grammar compared to our earlier examples. You may be wondering if there is a simpler *type* of grammar, perhaps a *context free grammar*, that can do the same job. In fact there is not. However, while the grammar is comparatively complex, the method it embodies in the generation of the sentences is quite simple. Essentially, like all grammars, it simply replaces one string by another at each stage in the derivation.

An interesting way of thinking about G_4 is in terms of it performing a kind of *computation*. Once a sentential form like $A^i B^j C$ is reached, the productions then ensure that $i \times j$ cs are appended to the end by essentially modelling the simple algorithm in Table 2.15.

The question that arises is: what range of computational tasks can we carry out using such purely syntactic transformations? We see from our example that the type 0 grammar simply specifies string substitutions. If we take our strings of as and bs as representing numbers, so that, say, a^6 represents the *number* 6, we see that G_4 is essentially a model of a process for multiplying together two arbitrary length numbers.

Later in this book, we encounter an abstract machine, called a *Turing machine*, that specifies string operations, each operation involving the replacing of only one symbol by another, and we see that the machine is actually as powerful as the type 0 grammars. Indeed, the machine is capable of performing a wider range of computational tasks than even the most powerful real computer.

However, we will not concern ourselves with these issues until later. In the next chapter, we encounter more of the fundamental concepts of formal languages: *syntax*, *semantics* and *ambiguity*.

Table 2.15 The "multiplication" algorithm embodied in grammar G_4.

for each A do
    for each B do
        put a c at the end of the sentential form
    endfor
endfor

EXERCISES

For exercises marked "†", solutions, partial solutions, or hints to get you started appear in "Solutions to Selected Exercises" at the end of the book.

2.1. Classify the following grammars according to the Chomsky hierarchy. In all cases, briefly justify your answer.

(a) †

$$S \rightarrow aA$$
$$A \rightarrow aS \mid aB$$
$$B \rightarrow bC$$
$$C \rightarrow bD$$
$$D \rightarrow b \mid bB$$

(b) †

$$S \rightarrow aS \mid aAbb$$
$$A \rightarrow \varepsilon \mid aAbb$$

(c)

$$S \rightarrow XYZ \mid aB$$
$$B \rightarrow PQ \mid S$$
$$Z \rightarrow aS$$

(d)

$$S \rightarrow \varepsilon$$

2.2.† Construct set definitions of each of the languages generated by the four grammars in exercise 1.

Hint: the language generated by 1(c) is not the same as that generated by 1(d), as one of them contains no strings at all, whereas the other contains exactly one string.

2.3.† It was pointed out above that we usually insist that one or more non-terminals must be included in the left-hand side of type 0 productions. Write down a formal expression representing this constraint. Assume that N is the set of non-terminals, and T the set of terminals.

2.4. Construct regular grammars, G_v, G_w and G_x, such that

(a) $L(G_v) = \{c^j : j > 0,$ and j does not divide exactly by 3$\}$

(b) $L(G_w) = \{a^i b^j [cd]^k : i, k \geq 0, 0 \leq j \leq 1\}$

> *Note: as we are dealing only with whole numbers, the expression* $0 \leq j \leq 1$, *which is short for* $0 \leq j$ *and* $j \leq 1$, *is the same as writing:* $j = 0$ *or* $j = 1$.

(c) $L(G_x) = \{a,\, b,\, c\}^*$

2.5.[†] Use your answer to exercise 4(c) as the basis for sketching out an intuitive justification that A^* is a regular language, for any alphabet, A.

2.6. Use the symbol $\Longrightarrow$ in showing the step-by-step derivation of the string c^5 using

(a) G_v
and
(b) G_x from exercise 4.

2.7. Construct context free grammars, G_y and G_z, such that

(a) $L(G_y) = \{a^{2i+1} c^j b^{2i+1} : i \geq 0,\, 0 \leq j \leq 1\}$
   Note: if $i \geq 0$, a^{2i+1} *means " all odd numbers of as".*

(b)[†] $L(G_z) = $ all Boolean expressions in your favourite programming language. (Boolean expressions are those that use logical operators such as "and", "or" and "not", and evaluate to *true* or *false*.)

2.8. Use the symbol $\Longrightarrow$ in showing the step-by-step derivation of $a^3 b^3$ using

(a) G_y from exercise 7, and the grammar

(b) G_3 from Chapter 2, i.e. $S \rightarrow ab \mid aSb$

2.9. Provide a regular grammar to generate the language $\{ab,\, abc,\, cd\}$.
   Hint: make sure your grammar generates only *the three given strings, and no others.*

210.[†] Use your answer to exercise 9 as the basis for sketching out an intuitive justification that any finite language is regular.
   Note that the converse of the above statement, i.e. that every regular language is finite, is certainly not *true. To appreciate this, consider the languages specified in exercise 4. All three languages are both regular and infinite.*

3
Syntax, Semantics and Ambiguity

3.1 Overview

In this chapter, we consider the following aspects of formal languages, and their particular relevance to programming languages:

- the *syntax*, or *grammatical structure*, of a sentence

- the *semantics*, or *meaning*, of a sentence

- the graphical representation of *derivations* by structures called *derivation* trees

- *parsing*, or trying to discover the grammatical structure of a given sentence

- *ambiguity*, when a sentence in a formal language has more than one possible meaning.

3.2 Syntax vs. Semantics

The *Phrase structure grammars* (*PSGs*), as introduced in the preceding chapter, describe the *syntax* of languages. The syntax of a language is the set of rules by which "well formed" phrases, which we call *sentences*, come about. The BNF definitions or syntax charts for Pascal to which we referred in the previous chapter tell us about the correct *form* of Pascal statements, but do not tell us what the statements cause the machine to do, after they have been compiled. They do not

A.P. Parkes, *A Concise Introduction to Languages and Machines*,
DOI: 10.1007/978-1-84800-121-3_3, © Springer-Verlag London Limited 2008

tell us how to write a program to compute a particular function, or solve a given problem. To write programs to do these things requires us to know about the *meaning* of the program constructs, i.e., their *semantics*.

Some of the languages we looked at in Chapter 2 had no semantics whatsoever (or at least none that we referred to). For semantics is not form alone, but also "interpretation", and, like syntax, requires that we have access to a set of rules which tell us how to make this interpretation. For natural languages such as English these rules of interpretation are extremely complex and not completely understood. Moreover, the rules of natural languages are not necessarily universal, and are subject to constant revision, especially in artistic usage.

The semantics of a sentence is its meaning. For a program, the term "semantics" refers to the computation carried out after the program source code has been compiled, when the program actually runs. In *formal languages*, meaning is inextricably linked to form. The grammatical structure of a sentence, by which we mean an account of how the *productions* were applied to obtain the sentence, is assumed to determine the meaning that we attribute to that sentence. We shall see certain implications of this below, but first we provide a foundation for our discussion by investigating a very useful way of representing the structure of context free *derivations*.

3.3 Derivation Trees

As we know, *context free grammars* have productions where every left-hand side consists of a single *non-terminal*, a property also shared by the *regular grammars*. In each step (application of one production) of the derivation, therefore, a single non-terminal in the *sentential form* will have been replaced by a string of zero or more *terminals* and/or non-terminals to yield the next sentential form. Any derivation of a sentential form begins with S, the *start symbol*, alone. These features mean that we can represent the derivation of any sentential form as a *tree*, and a very useful feature of trees is that they can be represented graphically, and therefore be used to express the derivation in a visually useful way. Trees used to represent derivations are called *derivation trees* (or sometimes *parse trees*, as we see later).

Let us describe how *derivation trees* are constructed by using examples. First, a simple tree, featuring grammar G_3 from Chapter 2:

$$S \rightarrow ab \mid aSb.$$

Consider the derivation:

$$S \Rightarrow aSb \Rightarrow aaSbb \Rightarrow aaabbb.$$

This can be represented as the tree in Figure 3.1.

Figure 3.2 shows how the derivation tree relates to the derivation itself.

In a tree, the circles are called *nodes*. The nodes that have no nodes attached beneath them are called *leaf* nodes, or sometimes *terminal nodes*. In such a tree, the resulting string is taken by reading the *leaf* nodes in left to right order.

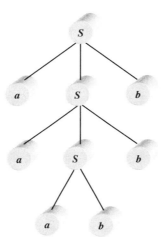

Figure 3.1 A derivation tree.

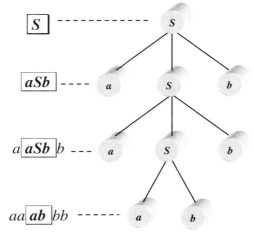

Figure 3.2 A derivation and its correspondence with the derivation tree.

A slightly more complex example is based on grammar G_2, also from Chapter 2:

$$S \rightarrow aB \mid bA \mid \varepsilon$$
$$A \rightarrow aS \mid bAA$$
$$B \rightarrow bS \mid aBB.$$

The derivation

$S \Rightarrow aB \Rightarrow aaBB \Rightarrow aabSB \Rightarrow aabB \Rightarrow aabaBB \Rightarrow aababSB \Rightarrow aababB \Rightarrow$
$aababbS \Rightarrow aababb$

is represented by the tree in Figure 3.3.

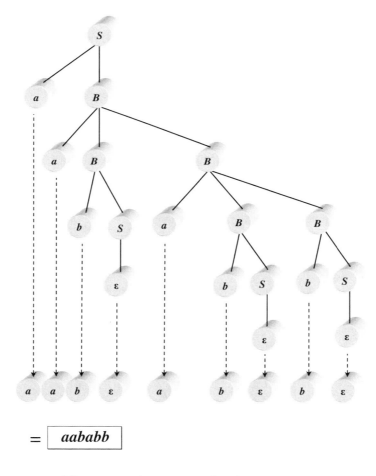

Figure 3.3 A more complex derivation tree.

Dashed lines have been included in Figure 3.3 to clarify the order in which we read the terminal symbols from the tree.

Note how, for the purposes of drawing the derivation tree, the *empty string*, ε, is treated exactly as any other symbol. However, remember that the final string, *aababb*, does not show the εs, as they disappear when *concatenated* into the resulting string.

3.4 Parsing

The *derivation* of a string, as described in Chapter 2, is the process of applying various productions to produce that string. In the immediately preceding example, we *derived* the string *aababb* using the productions of grammar G_2. *Parsing*, on the other hand, is the creation of a structural account of the string according to a grammar. The term *parsing* relates to the Latin for the phrase *parts of speech*. As for *derivation trees*, we also have *parse trees*, and the *derivation* trees shown above are also *parse* trees for the corresponding strings (i.e. they provide an account of the *grammatical structur*e of the strings).

Parsing is an important part of (formal and natural) language understanding. For the compilation of source programs it is absolutely crucial, since unless the compiler can arrive at an appropriate parse of a statement, it cannot be expected to produce the appropriate object code.

For now we look at the two overall approaches to parsing, these being *top-down* and *bottom-up*. The treatment given here to this subject is purely abstract and takes a somewhat extreme position. There are actually many different approaches to parsing and it is common to find methods that embody elements of both top-down and bottom-up approaches.

The essence of parsing is that we start with a grammar, G, and a *terminal string*, x, and we say:

- construct a derivation (parse) tree for x that can be produced by the grammar G.

Now, while x is a *terminal string*, it may not be a *sentence*, so we may need to precede the above instruction with "*find out if x is a sentence and if so ...*". As previously stated, there are two overall approaches to parsing, these being *top-down* and *bottom-up*. We now briefly describe the two approaches.

In *top-down parsing*, we build the parse tree for the terminal string x, starting from the top, i.e. from S, the start symbol. The most purely top-down approach would be to use the grammar to repeatedly generate different sentences until x was produced. *Bottom-up parsing* begins from the terminal string x itself, trying to build the parse tree from the sentence upwards, finishing when the tree is fully formed, with S, the start symbol, at the top.

To clarify bottom-up parsing, we will consider one bottom-up approach, known as the *reduction* method of parsing. We begin with our grammar, G, and make a grammar G_{red}, by swapping over the left and right-hand sides of all of the productions in G. We call the resulting "productions" *reductions*. The goal then becomes to use the *reductions* to eventually arrive at S after starting with the string x.

We again consider grammar G_3, i.e.

$$S \rightarrow aSb \,|\, ab$$

and the problem of producing the parse tree for $aaabbb$. The "reductions" version of G_3, called G_{3red}, is as follows:

$$aSb \rightarrow S$$

$$ab \rightarrow S.$$

We start with our sentence x, and seek a left-hand side of one of the reductions that matches some *substring* of x. We replace that substring by the right-hand side of the chosen reduction, and so on. We terminate when we reach a string consisting only of S. Parsing is simple in our example, which is shown in Figure 3.4, since only one reduction will be applicable at each stage.

In the example in Figure 3.4, there was never a point at which we had to make a choice between several applicable reductions. In general, however, a grammar may be such that at many points in the process there may be several applicable reductions, and several substrings within the current string that match the left-hand sides of various reductions.

As a more complex example, consider using the reduction method to produce a parse tree for the string $aababb$. This is a sentence we derived using grammar G_2 above. For G_2, the reductions G_{2red} are:

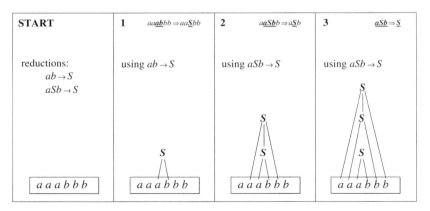

Figure 3.4 Reduction parsing.

$$\varepsilon \to S$$
$$aB \to S$$
$$bA \to S$$
$$aS \to A$$
$$bAA \to A$$
$$bS \to B$$
$$aBB \to B.$$

We have to begin with the $\varepsilon \to S$ reduction, to place an S somewhere in our string so that one of the other reductions will apply. However, if we place the S after the first a, for example, we will be unable to complete the tree (you should convince yourself of this). The same applies if we place the S after the second a in the string. In fact, you will appreciate that a reduction rule with an ε on its left-hand side can be applied *at any stage* in the parse. Moreover, suppose that later on in the parse we reach, say, $aabSaBB$. How do we choose between $aBB \to B$ and $aB \to S$?

We shall see in Chapters 4 and 5 how some of the problems of parsing can be simplified for certain classes of grammars, i.e. the *regular* grammars and (some of) the *context free* grammars. We do this by establishing that, associated with these grammars, are *abstract machines*. We then see that such machines can be used as a basis for designing parsing programs for the languages in question. For now, however, we look at an extremely important concept that relates to languages in general.

3.5 Ambiguity

Now that we have familiarised ourselves with the notion, and some of the problems, of parsing, we are going to consider one of the most important related concepts of formal languages, i.e. that of *ambiguity*. Natural language permits, and indeed in some respects actually thrives on, ambiguity. In talking of natural language, we usually say a statement is ambiguous if it has more than one possible meaning. The somewhat contrived example often used to demonstrate this is the phrase:

"Fruit flies like a banana."

Does this mean "the insects called fruit flies are positively disposed towards bananas"? Or does it mean "something called fruit is capable of the same type of trajectory as a banana"? These two potential meanings are partly based on the (at least) two ways in which the phrase can be *parsed*. The former meaning assumes that "fruit flies" is a single noun, while the latter assumes that the same phrase is a noun followed by a verb.

The point of the above example is to show that while our intuitive description of what constitutes ambiguity makes reference to *meaning*, the different meanings can in fact be regarded as the *implication* of the ambiguity. Formally, ambiguity in a language is reflected in the existence of more than one parse tree for one or more sentences of that language. To illustrate this, let us consider an example that relates to a programming language called Algol60, a language that influenced the designers of Pascal. The example also applies to Pascal itself and is based on the following fragment of a programming language definition, for which we revert, for convenience, to the *BNF* notation introduced in Chapter 2.

<statement> ::= <if statement> | <arithmeticexpression> | ...

<if statement> ::= *if* <Boolean expression> *then* <statement> |

if <Boolean expression> *then* <statement> *else*

<statement>

Remember that the terms in angled brackets ("<statement>", etc.) are *non-terminals*, and that other items (*if, then* and *else*) are regarded as *terminals*.

Now consider the following statement:

if x > y then if y < z then x := x + 1 else x := x − 1.

Suppose that x > y and y < z are <Boolean expression>s, and that x := x + 1 and x := x − 1 are <arithmetic expression>s. Then two different parse trees can be constructed for our statement. The first is shown in Figure 3.5 and is labelled "PARSE TREE 1".

"PARSE TREE 2", an alternative parse tree for the same statement is shown in Figure 3.6.

We have two distinct structural accounts of a single sentence. This tells us that the grammar is *ambiguous*.

Now, suppose that the compiler for our language used the structure of the parse tree to indicate the order in which the parts of the statement were executed. Let us write out the statement again, but this time indicate (by inserting "[" and "]" into the statement) the interpretation suggested by the structure of each of the two trees.

PARSE TREE 1 (Figure 3.5) suggests

if x > y then [if y < z then x := x + 1 else x := x − 1].

However, PARSE TREE 2 (Figure 3.6) suggests

if x > y then [if y < z then x := x + 1] else x := x − 1.

The "else part" belongs to different "if parts" in the two interpretations. This would clearly yield different results according to which interpretation our

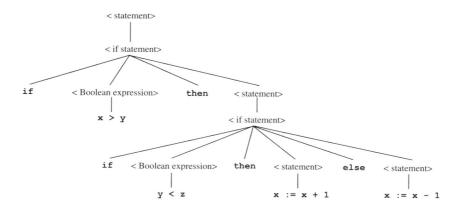

Figure 3.5 PARSE TREE 1: A derivation tree for a Pascal "if statement".

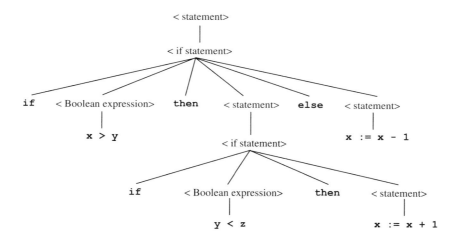

Figure 3.6 PARSE TREE 2: A different derivation tree for the same "if statement" as that in Figure 3.5.

compiler made. For example, if $x = 1$, $y = 1$ and $z = 2$, then in case 1 the execution of our statement would result in x still having the value 1, while after execution in case 2, x would have the value 0.

More seriously, a compiler for our language on one manufacturer's machine may create PARSE TREE 1 (Figure 3.5), while a compiler for another type of machine creates PARSE TREE 2 (Figure 3.6). What is more, a note would probably have to be added to the syntax descriptions of our programming language, e.g. in the manual, to explain which interpretation was assumed by the particular compiler. This is not a good policy to adopt in programming language definition, as the syntax for the programming language ought to be specified in a way that is subject to a single interpretation, and that interpretation should be obvious from looking at the formal definition of the syntax. That is to say, the language should be *unambiguous*.

Now, our example statement above is actually syntactically correct Pascal, as an "if" statement is not a compound statement in Pascal, and therefore the subordinate if statement ("if y < z then ...") does not need to be bracketed with begin and end. Thus, there is ambiguity in the standard definition of Pascal. The solution to this in the original Pascal manual was to inform the reader (in an additional note) which interpretation would be taken. The solution adopted for Algol was to change a later version of the language (Algol68) by introducing "bracketing tokens"; sometimes called "guards" (similar to the *endif*s used in the algorithms in this book). If we have bracketing tokens and we would like the interpretation specified by PARSE TREE 1 (Figure 3.5), we write:

if x > y then if y < z then x := x + 1 else x := x − 1 endif endif .

If, on the other hand, we desire the PARSE TREE 2 (Figure 3.6) interpretation we write:

if x > y then if y > z then x := x + 1 endif else x := x − 1 endif .

The difference in meaning between the two statements is now very clear.

While ambiguity is, as mentioned above, an accepted part of natural languages, it *cannot* be accepted in programming languages, where the consequences of ambiguity can be serious. However, ambiguity in natural language also leads to problems on occasions. Krushchev, a president of the former Soviet Union, spoke a phrase to western politicians that was translated into English literally as "Communism will bury you." This was interpreted to be a threat that the Soviet Union would destroy the west, but in Russia the original phrase also has the meaning "Communism will outlast you!"

To complete this introduction to ambiguity, we will precisely state the definition of ambiguity in formal grammars.

A Phrase Structure Grammar (PSG), G, is *ambiguous* if $L(G)$ contains *any* sentence which has two or more distinct derivation trees.

Correspondingly:

A PSG, G, is *unambiguous* if $L(G)$ contains *no* sentences which have two or more distinct derivation trees.

If we find any sentence in $L(G)$ which has two or more derivation trees, we have established that our grammar is ambiguous. Sometimes, an ambiguous grammar can be replaced by an *un*ambiguous grammar which does the same job (i.e. generates exactly the same language). This was not done to solve the Algol60 problem described above, since in that case new terminal symbols were introduced (the bracketing terms), and so the new grammar generated many different sentences from the original grammar, and therefore the *language* was changed.

Ambiguity is problematic, as there is no general solution to the problem of determining whether or not an arbitrary PSG is ambiguous. There are some individual cases where we can show that a grammar is unambiguous, and there are some cases where we can show that a given grammar is ambiguous (exactly what happened in the case of Algol60). However, there is no general solution to the ambiguity problem, since establishing that a grammar is unambiguous could require that we make sure that there is absolutely no sentence that the grammar generates for which we can construct more than one derivation tree.

In programming languages, ambiguity is a great cause for concern, as stated above. The main reason for this is that the compiler is usually designed to use the parse tree of a program to derive the structure of the object code that is generated. It is obvious therefore, that the programmer must be able to write his or her program in the knowledge that there is only one possible interpretation of that program. Being unable to predict the behaviour of a statement in a program directly from the text of that statement could result in dire consequences both for the programmer and the user.

EXERCISES

For exercises marked "†", solutions, partial solutions, or hints to get you started appear in "Solutions to Selected Exercises" at the end of the book.

3.1. Given the following grammar, G:

$$S \to XC \,|\, AY$$
$$X \to aXb \,|\, ab$$
$$Y \to bYc \,|\, bc$$
$$A \to a \,|\, aA$$
$$C \to c \,|\, cC.$$

(a)   Classify G according to the *Chomsky* hierarchy.

(b)†   Write a *set definition* of $L(G)$.

(c)[†]   Using the sentence $a^3b^3c^3$ in $L(G)$, show that G is an *ambiguous* grammar.

Note: $L(G)$ *is known as an "inherently ambiguous language", as any context free grammar that generates it is necessarily ambiguous. You might like to justify this for yourself (hint: think about the derivation of sentences of the form $a^ib^ic^i$).*

3.2. Given the following productions of a grammar, G:

$$S \to E$$
$$E \to T \mid E + T \mid T - E$$
$$T \to 1 \mid 2 \mid 3$$

(note that $1, 2, 3, +$and - are *terminal* symbols);

and the following *sentence* in $L(G)$:

3–2+1

(a)   use the sentence to show that G is an *ambiguous grammar*

(b)[†]   assuming the standard arithmetic interpretation of the terminal symbols, and with particular reference to the example sentence, discuss the *semantic* implications of the ambiguity.

3.3. Discuss the semantic implications of the ambiguity in the grammar:

$$P \to P \text{ or } P \mid P \text{ and } P \mid x \mid y \mid z$$

(where "or" and "and" are terminals),

assuming the usual Boolean logic interpretation of "or" and "and". Introduce new terminals "(" and ")" in such a way to produce a grammar that generates similar, but unambiguous logical expressions.

3.4. Given the following ambiguous grammar:

$$S \to T \mid A \mid T \text{ or } S \mid A \text{ or } S$$
$$A \to T \text{ and } T \mid T \text{ and } A \mid \text{not } A$$
$$T \to a \mid b \mid c \mid \text{not } T$$

(where *or, and, not, a, b* and *c* are *symbols*).

Discuss possible problems that the ambiguity might cause when interpreting the sentence *a or not b and c* (assuming the usual Boolean logic interpretation of the symbols *and, or* and *not*).

4
Regular Languages and Finite State Recognisers

4.1 Overview

This chapter is concerned with the most restricted class of languages in the *Chomsky hierarchy* (as introduced in Chapter 2): the *regular* languages. In particular, we encounter our first, and most simple, type of abstract machine, the *Finite State Recogniser (FSR)*.

We discover:

- how to convert a *regular grammar* into an FSR

- how any FSR can be made *deterministic* so it never has to make a choice when *parsing* a regular language

- how a *deterministic FSR* converts directly into a simple *decision program*

- how to make an FSR as small as possible

- the *limitations* of the regular languages.

4.2 Regular Grammars

As we saw in Chapter 2, a *regular grammar* is one in which every production conforms to one of the following patterns:

A.P. Parkes, *A Concise Introduction to Languages and Machines*,
DOI: 10.1007/978-1-84800-121-3_4, © Springer-Verlag London Limited 2008

$X \rightarrow xY$ where X and Y are each *single non-terminals*, x is a

$X \rightarrow y$ terminal, and y is either the *empty string* (ε), or a *single terminal*.

Here are some productions that all conform to this specification (A and B are non-terminals, a and b are terminals):

$$A \rightarrow \varepsilon$$
$$A \rightarrow aB$$
$$B \rightarrow bB$$
$$A \rightarrow bA$$
$$A \rightarrow b$$
$$B \rightarrow a.$$

When we discussed *reduction parsing* in Chapter 3, we saw that productions such as $A \rightarrow \varepsilon$, with ε, the empty string on the right-hand side can cause problems in parsing. However, for reasons to be discussed later, we can ignore them for the present and assume for now that regular grammars contain no productions with ε on the right-hand side.

Here is an example of (the productions of) a regular grammar, G_5:

$$S \rightarrow aS \,|\, aA \,|\, bB \,|\, bC$$
$$A \rightarrow aC$$
$$B \rightarrow aC$$
$$C \rightarrow a \,|\, aC$$

and here is an example *derivation* of a *sentence*:

$$S \Rightarrow aS \Rightarrow aaS \Rightarrow aaaA \Rightarrow aaaaC \Rightarrow aaaaa.$$

An equivalent *derivation tree* as introduced in Chapter 3, can be seen in Figure 4.1.

You may have observed that the above grammar is also *ambiguous* (*ambiguity* was discussed in Chapter 3), since there is at least one alternative derivation tree for the given sentence. Try to construct one, to practise creating derivation trees.

In this chapter, we use G_5 to illustrate some important features of regular languages. For now, you might like to try to write a *set definition* of the language generated by the grammar. You can assess your attempt later in the chapter, when $L(G_5)$ is defined.

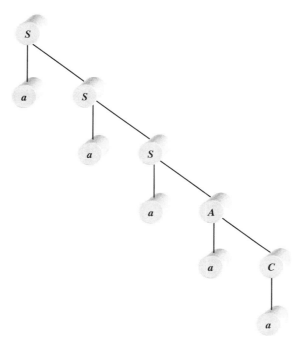

Figure 4.1 A derivation tree produced by a regular grammar.

4.3 Some Problems with Grammars

It is not always straightforward to define a language by examining the corresponding grammar. Take our example grammar G_5 above. One problem is that various symbols representing the same entity are scattered throughout the productions (the non-terminal C, for example). Another problem is that, as computer scientists, we are not usually satisfied by merely designing and writing down a grammar, we usually want to write a *parser* for the language it generates. This leads to questions that grammars do not conveniently allow us to address. Will our grammar yield an efficient parser (in terms of space and time)? If we design another grammar with fewer productions, say, to generate the same language, is there a way of being sure that it does indeed do this? What about the design of grammars themselves: is there a more convenient design notation which we can use in place of the textual form? Then there are productions, such as $S \to aS$ and $S \to aA$, that may cause problems in the parsing of the language, as was discussed in Chapter 3. Can we eliminate situations where there is a choice, such as these?

For the regular grammars, the answer to all of the above questions is "yes".

4.4 Finite State Recognisers and Finite State Generators

For the regular grammars there is a corresponding *abstract machine* that can achieve the same tasks (parsing and generating) as those supported by the grammar. Moreover, as the structure of the regular grammar is so simple, the corresponding machine is also simple. We call the machine the *finite state recogniser* (*FSR*). It is called a *recogniser* because it "recognises" if a string presented to it as input belongs to a given language. If we were using the machine to *produce*, rather than *analyse*, strings, we would call it a *finite state generator*.

4.4.1   Creating an FSR

We can devise an *equivalent* FSR from any regular grammar, by following the simple rules in Table 4.1.

Figure 4.2 demonstrates the application of part (a) of the instructions in Table 4.1 to our example grammar G_5.

Table 4.1 Constructing an FSR from the productions of a regular grammar.

The construction rules (X and Y denote any non-terminals, y denotes any terminal):

(a) Convert all productions using whichever of the following three rules applies, in each case:

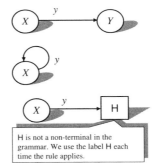

    1. a production of the form
        $X \rightarrow yY (Y \neq X)$
    becomes a drawing of the form

    2. a production of the form
        $X \rightarrow yX$
    becomes a drawing of the form

    3. a production of the form
        $X \rightarrow y$
    becomes a drawing of the form:

                                 H is not a non-terminal in the
                                 grammar. We use the label H each
                                 time the rule applies.

(b) Take all the drawings produced in (a), and join all of the circles and squares with the same labels together.
    (Once the basic concept is appreciated, one usually finds the whole machine can be drawn directly.)

(c) The circle (S) (where S is the *start symbol* of the grammar) becomes (S)
    (i.e. an arrow is added that comes from nowhere and points to S).

The result is a pictorial representation of a *directed graph* ("*digraph*") that we call a *finite state recogniser* (*FSR*).

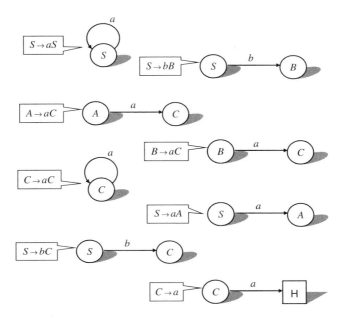

Figure 4.2 The first part of expressing a regular grammar as an FSR. This represents the application to grammar G_5 of part (a) of Table 4.1.

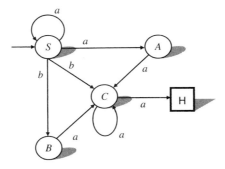

Figure 4.3 The FSR M_5 that represents regular grammar G_5.

Figure 4.3 shows the machine that is produced after parts (b) and (c) of the procedure in Table 4.1 have been carried out.

To show its correspondence with the grammar from which it was constructed (G_5), we will call our new machine M_5.

To support a discussion of how our machines operate, we need some new terminology, which is provided by Table 4.2.

Table 4.2 Terminology for referring to FSRs.

Term	Pictorial representation (from M_5 – Figure 4.3)
states (also called *nodes*)	S A B C H
arcs Note: this is a *labelled* arc. The label is the terminal a. The arc that enters state S from nowhere is an *unlabelled* arc.	$X_1 \xrightarrow{\;a\;} X_2$
ingoing arcs	C
outgoing arcs	C
Special states *start state* (has an *ingoing* arc coming from nowhere)	$\rightarrow S$
halt or *acceptance state*	H

4.4.2 The Behaviour of the FSR

A machine is of little use if it does not do something. Our machines are abstract, so we need to specify their rules of operation. Table 4.3 provides us with these rules.

It can be seen from Table 4.3 that there are two possible conditions that cause the halting of the machine:

(1) the input is *exhausted*, so no transitions can be made

(2) the input is *not exhausted*, but no transitions are possible.

It can also be seen that one of two possible additional conditions must prevail when the machine halts:

(A) it is in an *acceptance* state

(B) it is in a *non-acceptance* state.

Our machine is very simple and does not produce output as a program might, and so, if we imagine that we are blessed with the power to observe the internal workings of the machine, the situation that we observe at the time of halting is of

Table 4.3 The rules governing the behaviour of the FSR.

Rule №	Description
1	The FSR begins its operation from any of its *start states*. It is said to be *in* that start state.
	It has available to it any string of terminal symbols, called the *input string*. If the input string is non-empty, its leftmost symbol is called the *current input symbol*.
2	When in any state, q (the symbol q is often used to denote states in abstract machines) it can make a move, called a *transition* (or *state transition*) to any other (possibly the same) single state, q_1, if and only if there is an arc as follows (x represents the *current input symbol*):

Note that either or both of q or q_1 can be halt states.

After making a transition, the following conditions hold:

- the machine is *in* state q_1
- the symbol in the input string immediately to the right of the current input symbol now becomes the current input symbol. If there is no such symbol we say that the input is *exhausted*.

	If, at any point, any transition is possible, the machine *must* make one of them. However, at any point where more than one transition is possible, the choice between the applicable transitions is completely arbitrary.
	Note: when the machine makes a transition and moves to the next symbol in the input string, we say that the machine is *reading* the input.
3	When the machine is in a given state, and can make no transition whatsoever, the machine is said to have *halted* in that state.

Table 4.4 Possible state × input string configurations when the FSR halts.

	Condition of input	
State in which FSR halts	input exhausted	input not exhausted
acceptance (halt) state	accepted	rejected
non-acceptance state	rejected	rejected

Table 4.5 String acceptance and rejection conditions for the FSR.

When the FSR halts ...

... if the FSR is in an *acceptance* (*halt*) state and the input is *exhausted,* we say that the input string is *accepted* by the FSR

... if the FSR is not in an *acceptance* (*halt*) state, or is in a halt state and the input is not *exhausted,* we say that the input string is *rejected* by the FSR.

critical importance. What we will observe when the machine stops is the situation "inside" the machine with respect to the combination of (1) or (2) with (A) or (B) from above, as represented in Table 4.4.

Table 4.5 defines the *acceptance* and *rejection* conditions of the FSR.

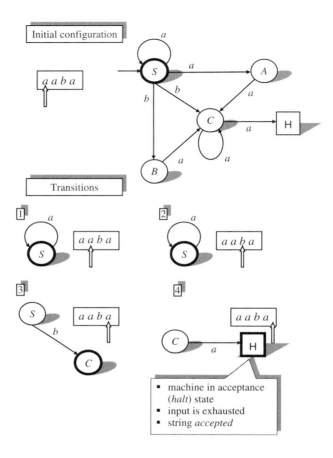

Figure 4.4 M_5 accepts the string $aaba$.

We now consider examples of the machine M_5 in operation. These examples are shown in Figures 4.4 to 4.6. In each case, we show the initial configuration of the machine and input, followed by the sequence of transitions. In each of Figures 4.4 to 4.6, the emboldened state in a transition represents the state that the machine reaches on making that transition. The arrow beneath the input string shows the new current input symbol, also after a transition is made.

Example 1. Input string: $aaba$.

Figure 4.4 shows how the string $aaba$ can be *accepted* by M_5. You can observe that several choices were made, as if by magic, as to which transition to take in a given situation. In fact, you could show a sequence of transitions that lead

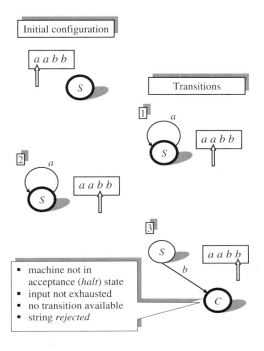

Figure 4.5 M_5 rejects the string *aabb*.

to *aaba* being *rejected*. This appears to be a problem with rule 2 in Table 4.3, to which we return later. First, another example.

Example 2. Input string: *aabb*.

In Figure 4.5, we see the string *aabb* being *rejected* by M_5. Although choices were made at several points, you can easily establish that *there is no sequence of transitions that will lead to the string aabb being accepted by M_5*.

The next example shows another string being rejected under different *halting* conditions.

Example 3. Input string: *aaabb*.

As in example 2 (Figure 4.5), you can see that no sequence of transitions can be found that would result in the input string in Figure 4.6 being accepted by M_5. The final combination of halting conditions (*input exhausted,not in acceptance state*) can be achieved for the string *aaba*, which we saw being accepted above (example 1). We see a sequence of transitions that will achieve these rejection conditions for the same string later. For the moment, you may wish to discover such a sequence for yourself.

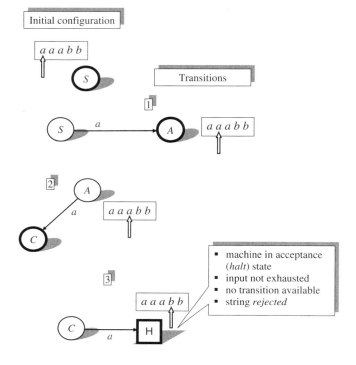

Figure 4.6 M_5 rejects the string *aaabb*.

4.4.3   The FSR as Equivalent to the Regular Grammar

Let us remind ourselves why we created the machine in the first place. Recall the earlier statement about the machine being *equivalent* to the original grammar. What this ought to mean is that all strings accepted by the machine are in the language generated by the grammar (and *vice versa*). The problem is that, according to the particular transition chosen in the application of rule 2 (Table 4.3), the machine can sometimes reject a string that (like *aaba* in example 1, above) *is* generated by the grammar. We want to say something such as: "all strings that *can* be accepted by the machine are in the language generated by the grammar, and all strings that are generated by the grammar *can* be accepted by the machine". That is exactly what we do next.

First we say this:

- a string is *acceptable* if there is some sequence of *transitions* which lead to that string being *accepted*.

Thus, *aaba*, from example 1, is acceptable, even though we could provide an alternative sequence of transitions that would lead to the string being rejected. On the other hand, the strings from examples 2 and 3 are not acceptable and also cannot be generated by the grammar. You may wish to convince yourself that the preceding statement is true.

Suppose that G_z is any regular grammar, and we call the finite state recogniser derived from G_z's productions M_z; we can say that *the set of acceptable strings of Mz is exactly the same as* $L(G_z)$, *the language generated by the grammar* G_z. The rules we followed to produce M_z from G_z ensure that this is the case, but to convince ourselves, let us argue the case thoroughly.

To prove that two sets (in this case, *the set of acceptable strings of Mz* and $L(G_z)$) are equal, we have to show that all members of one set are members of the other, and *vice versa*. If we carry out the first part only, we are merely showing that the first set is a *subset* of the second. We need to show that *each set is a subset of the other*. Only then can we be sure that the two sets are equal.

In the following, L_z is an abbreviation for $L(G_z)$.

Part 1. All strings in L_z are acceptable to M_z.

Let x be any sentence of L_z, such that $|x| = n$, where $n \geq 1$, i.e. x is a non-empty string of length n. We can *index* the symbols of x, as discussed in Chapter 2, and we write it $x_1, x_2, ..., x_n$. We show that a derivation tree for x directly represents parts of the machine M_z. A derivation tree for x is of the form shown in Figure 4.7.

The derivation tree in Figure 4.7 can be transformed, in an obvious way, into the diagram shown in Figure 4.8.

We can do this because each part of the derivation tree like this:

used a production of the form $X \rightarrow xY$. The expansion of the lowest non-terminal node in the tree will look like this:

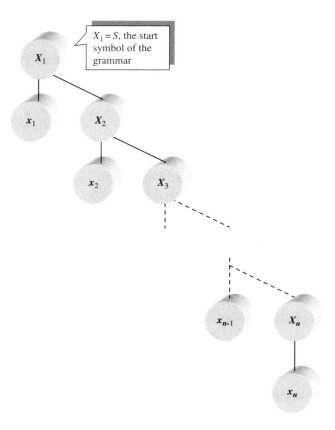

Figure 4.7 A derivation tree for an arbitrary regular grammar derivation.

which used a production of the form $X \rightarrow x$ (single non-terminal to single terminal). The sequence of productions used in Figure 4.7 become, if the conversion algorithm is applied, the FSR of Figure 4.8, since $X_1 = S$ (the start symbol of the grammar), and a production of the form $X \rightarrow x$ becomes an arc:

Some of the non-terminals may, of course, appear more than once in the derivation tree, and may have to be joined together (as specified in part (b) of Table 4.1). For any sentence, x, of the language there will thus be a path in the FSR that will accept that sentence – x is an *acceptable* string to M_z.

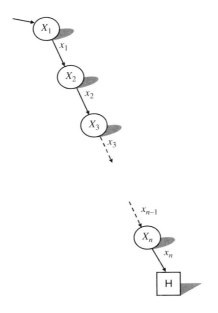

Figure 4.8 Conversion of the tree from Figure 4.7 into an FSR.

Part 2. All acceptable strings of M_z are in L_z.

Let x be any acceptable string of M_z. Choose any path through M_z that accepts x. We can write out this path so that it is has no *loops* in it (as in Figure 4.8). We then convert this version into a tree using the converse of the rules applied to the derivation tree in Figure 4.7. You can easily justify that the resulting tree will be a derivation tree for x that uses only productions of G_z. Thus, any *acceptable* string of M_z can be generated by G_z and is thus in L_z.

We have shown (part 1) that every string generated by G_z is acceptable to M_z, and (part 2) that every string acceptable to M_z is generated by G_z. We have thus established the *equivalence* of M_z and G_z.

4.5 Non-determinism in Finite State Recognisers

As was pointed out above, rule 2 of the rules governing the behaviour of an FSR (see Table 4.3) implies that there may be a choice of transitions to make when the machine is in a given state, with a given current input symbol. As we saw in the case of example 1, a wrong choice of transition in a given state may result in an acceptable string being rejected. It is not rule 2 itself that is at fault, it is the type of machine that we may produce from a given set of productions. We will now state this more formally.

- If a finite state recogniser/generator contains any state with two or more identically labelled *outgoing arcs*, we call it a *non-deterministic finite state recogniser*.

- If a finite state recogniser/generator contains no states with two or more identically labelled *outgoing arcs*, we call it a *deterministic finite state recogniser*.

It should be clear from this that M_5, our example machine (Figure 4.3), is *non-deterministic*. If a machine is deterministic, then in the process of accepting or rejecting a string, it *never has to make a choice as to which transition to take, for a given current input symbol, when in a given state; there is only one possible transition which can be made at that point*. This considerably simplifies the processes of parsing, as we shall see in the next section. Incidentally, we can still use the same rules for FSR behaviour as those above, for *deterministic* FSRs; the "choice" of transition will at each point consist of only one option.

In following our rules to produce an FSR from the productions of a regular grammar, we may create a non-deterministic machine. Remember our machine M_5 (Figure 4.3), and in particular our discussion about input string *aaba*. If the wrong sequence of transitions is chosen, for example:

$$S \text{ to } S \text{ reading an } a,$$
$$S \text{ to } S \text{ reading an } a,$$
$$S \text{ to } B \text{ reading a } b,$$
$$B \text{ to } C \text{ reading an } a,$$

the machine's halting conditions would be *input exhausted, not in halt state*; indicating a rejection of what we know to be an *acceptable* string. This implies that a parser that behaves in a similar way to a non-deterministic FSR might have to try many sequences of applicable transitions to establish whether or not a string was acceptable. Worse than this, for a non-acceptable string, the parser might have to try *all possible sequences of applicable transitions of length equal to the length of the string* before being sure that the string could indeed be rejected.

The main problem associated with non-determinism, therefore, is that it may result in parsers that have to *backtrack*. Backtracking means returning to a state of affairs (for FSRs this means to a given *state* and *input symbol*) at which a choice between applicable transitions was made, in order to try an alternative. Backtracking is undesirable (and as we shall see, for regular languages it is also *unnecessary*), since

- a backtracking parser would have to save arbitrarily long input strings in case it made the wrong choice at any of the choice points (as it may have made several transitions *since* the choice point),

and

- backtracking uses up a lot of time.

If an FSR is deterministic there is, in any given state, only one choice of transition for any particular input symbol, which means that a corresponding parser *never* has to backtrack.

4.5.1 Constructing Deterministic FSRs

It may surprise you to learn that *for every set of regular productions we can produce an equivalent deterministic FSR*. As we will see, we can then use that FSR to produce a simple, but efficient, deterministic decision program for the language. First, we consider a procedure to convert any non-deterministic FSR into an *equivalent* deterministic FSR. We will call the old machine M and the new deterministic machine M^d.

Table 4.6 presents the procedure, which is called the *subset construction algorithm*. Our machine M_5 will be used to illustrate the steps of the procedure. Table 4.6, followed by Figures 4.9 to 4.14 show how the new machine M_5^d is built up.

Continuing the execution of step 2, we will choose the state of M_5^d with A and S in it (see the note that follows step 2 of Table 4.6). This results in the partial machine shown in Figure 4.9.

We still have states of M_5^d with no *outgoing arcs* in Figure 4.9, so we choose one (in this case, the one with B and C inside it). The result is shown in Figure 4.10.

Now, from Figure 4.10 we choose the state containing A, S and C, resulting in Figure 4.11.

From Figure 4.11, we choose the state that contains A, S, C and H. The result can be seen in Figure 4.12.

Then we choose the state in Figure 4.12 that contains only C and H. Figure 4.13 shows the outcome.

Finally, from Figure 4.13 we choose the empty state of M_5^d. This state represents unrecognised input in certain states of M_5. Thus, if M_5^d finds itself in this state it is reading a string that is not in the language. To make M_5^d totally deterministic, we need to cater for the remainder (if any) of the string that caused M_5^d to reach this "*null*" state. Thus we draw an arc, for each terminal in M_5's alphabet, leaving and immediately re-entering the null state. The final machine is as shown in Figure 4.14.

Table 4.6 The subset algorithm.

Step	Description
1 (a)	Copy out all of the start states of M.
	Note: if M was constructed from a regular grammar it will have only one start state. However, in general, an FSR can have several start states (see later in this chapter).
(b)	If all of the states you just drew are non-acceptance states, draw a circle (or ellipse) around them all, otherwise draw a square (or rectangle) around them all. We call the resulting object a *state* of M^d.
(c)	Draw an *unlabelled ingoing* arc (see Table 4.2) to the shape you drew in (b).

Note: the result of step 1 for M_5 is:

2	Choose any state of M^d that has no outgoing arcs.
	for each terminal, t, in the alphabet of M do
	for each state, q, of M inside the chosen state of M^d do
	copy out each state of M which has an ingoing arc labelled t leaving q (but only copy out any given state once)
	endfor
	if you just copied out NO states, draw an empty circle else do step 1(b) then return here
	if the state of M^d you've just drawn contains exactly the same states of M as an existing state of M^d, rub out the state of M^d you've just drawn
	Draw an arc labelled t from the chosen state of M^d to either
	(i) the shape you just drew at (b)
	or
	(ii) the state of M^d which was the same as the one you rubbed out
	endfor

Note: the result of the first complete application of step 2 to M is:

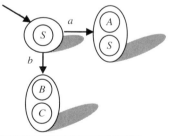

3	Repeat step 2 until every single state of M^d has exactly one outgoing arc for each terminal in M's alphabet.

If the machine ever finds itself in the null state, it will never leave it, but will continue to process the input string until the input is *exhausted*. It will not accept such a string, however, since the null state is not a halt state.

The subset algorithm now terminates (see step 3 of Table 4.6), leaving us with a totally deterministic machine, $M_5{}^d$ (shown in Figure 4.14) which is equivalent to M_5 (Figure 4.3).

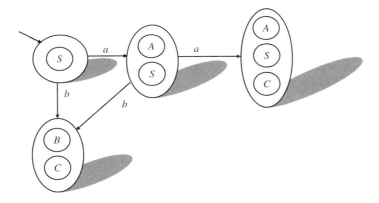

Figure 4.9 The subset algorithm in action (1).

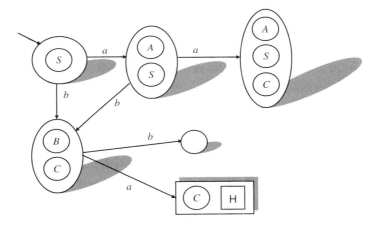

Figure 4.10 The subset algorithm in action (2).

We can now tidy up $M_5{}^d$ by renaming its states so that the names consist of single symbols, resulting in the machine we see in Figure 4.15.

This renaming of states illustrates that the names of states in an FSR are insignificant (as long as each label is distinct). What is important in this case is that there is a one-to-one association between the states and arcs of the old $M_5{}^d$ and the states and arcs of the renamed machine.

The null state and its ingoing and outgoing arcs are sometimes removed from the deterministic machine, their presence being assumed. Later, we will see that the null state in the deterministic FSR has its uses.

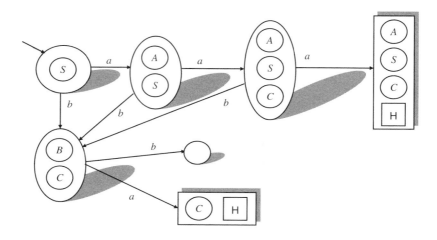

Figure 4.11 The subset algorithm in action (3).

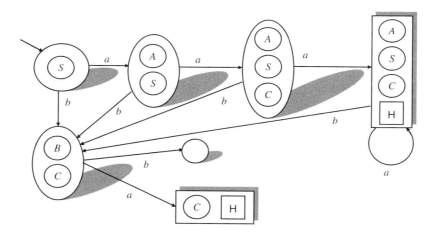

Figure 4.12 The subset algorithm in action (4).

4.5.2 The Deterministic FSR as Equivalent to the Non-deterministic FSR

We will now convince ourselves that M_5 and $M_5{}^d$ are equivalent. From $M_5{}^d$, we observe the following properties of acceptable strings:

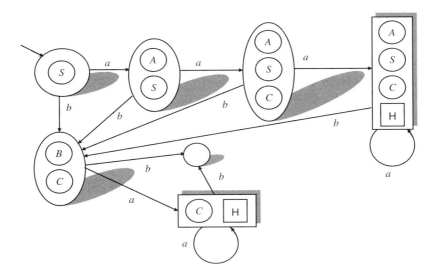

Figure 4.13 The subset algorithm in action (5).

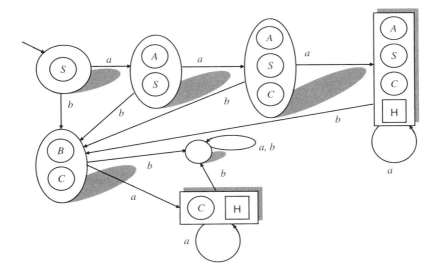

Figure 4.14 The result of the subset algorithm.

- if they consist of *a*s alone, then they consist of three or more *a*s

- they can consist of any number (≥ 0) of *a*s, followed by exactly one *b*, followed by one or more *a*s.

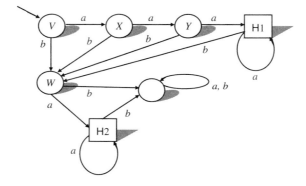

Figure 4.15 The FSR of Figure 4.14 with renamed states – $M_5{}^d$.

This can be described in set form:

$$\{a^i : i \geq 3\} \cup \{a^i b a^j : i \geq 0, j \geq 1\}$$

You can verify for yourself that all strings accepted by the original machine (M_5), of Figure 4.3, are also described by one of the above statements.

If we establish that for any non-deterministic FSR, M, we can produce an equivalent deterministic machine, M^d, we will, for the remainder of our discussion about regular languages, be able to assume that any FSR we discuss is deterministic. This is extremely useful, as we will see.

Suppose that M is any FSR, and that M^d is the machine produced when we applied our subset method to M. We will establish two things: (1) every string which is *acceptable* to M is *accepted* by M^d; (2) every string which is *accepted* by M^d is *acceptable* to M.[1]

Part 1. Every acceptable string of M is accepted by M^d.

Let x be an *acceptable* string of M. Suppose x has length n, where $n \geq 1$. We can *index* the symbols in x, as discussed in Chapter 2: $x_1, x_2, ..., x_n$. x is acceptable to M, so there is a path through M, from a start state, S, to a halt state, H, as represented in Figure 4.16.

---

[1] Note that this is not the only way of doing this: we could show (1) and then for (2) show that every string that is *non-acceptable* to M is *rejected* by M^d. You might like to try this approach.

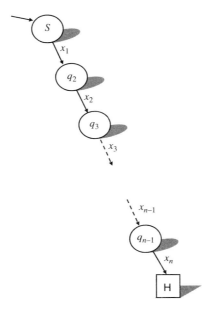

Figure 4.16 How a string is accepted by an FSR.

The states on the path in Figure 4.16 are not necessarily all distinct. Also, if $n = 1$, the path looks like this:

With respect to the above path through M, M^d will be as represented in Figure 4.17.

The string x, which was acceptable to M, is thus accepted by M^d. The above representation of the path taken by the string x through M^d may also incorporate any paths for the same string which result in the situations:

- *input exhausted, not in halt state* (the non-halt state of M reached will be incorporated into the halt state of M^d),

- *input not exhausted, in halt state* (other states of M^d on the path in Figure 4.17 may be halt states),

and

- *input not exhausted, not in halt state* (M would have stopped in one of the states contained in the non-halt states of Figure 4.17, but M^d will be able to proceed to its halt state with the remainder of the string).

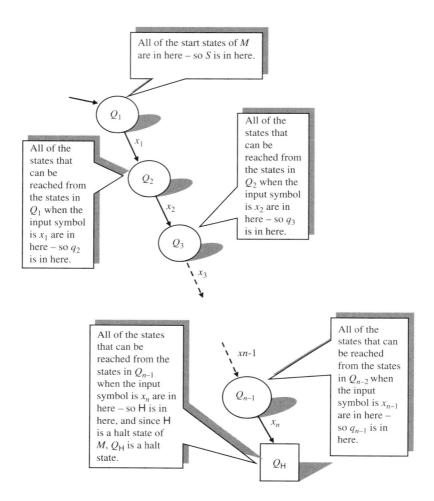

Figure 4.17 How a deterministic version of an FSR (M) accepts the string of Figure 4.16. The states Q_i represent the composite states produced by the subset algorithm. The states q_i represent the states of the original non-deterministic machine from Figure 4.16.

Part 2. Every string accepted by M^d is acceptable to M.

If a string, x, is accepted by M^d, the way the machine was constructed ensures that there is only one path through M^d for that string, as represented in Figure 4.17, if, again, we index the symbols of x as $x_1, x_2, ..., x_n$. Obviously there exist one or more paths in M that will accept the string x. I leave it to you to convince yourself of this.

We have seen (part 1) that every string accepted by M, the non-deterministic FSR, is accepted by M^d, the deterministic version of M. Moreover, we have seen (part 2) that every string accepted by M^d is acceptable to M. M and its deterministic counterpart, M^d, are *equivalent* FSRs.

4.6 A Simple Deterministic Decision Program

The FSR is not only a more convenient representation of the productions of a regular grammar in some respects, but, if deterministic, also represents a simple decision program for the corresponding language. It is straightforward to convert a deterministic FSR into an equivalent decision program. Here, we will produce a "pseudo-Pascal" version. A similar program was presented in Chapter 2; you may like to examine it again.

Figure 4.18 illustrates the conversion of a deterministic FSR into a decision program by associating parts of our deterministic machine $M_5{}^d$ with modules of the program. The example should be sufficient to enable you to infer the rules being used in the conversion.

Note that the use of quoted characters 'a' and 'b' in the program in Figure 4.18 is not meant to suggest that data types such as Pascal's "char" are generally suitable to represent terminal symbols. (A related discussion can be found in the "Solutions to Selected Exercises" section, in the comments concerning exercise 3 of this chapter.)

You should appreciate the importance of the *null state* to the construction of our parser: without it, our program may have had no defined action in one or more of the "case" statements. Whether or not the program continues to process the input string when it reaches the null state (the code segment labelled *NLL*), as does our program in Figure 4.18, is a matter of preference. In certain circumstances one may require the program to abort immediately (with a suitable message), as soon as this segment is entered, not bothering with any remaining symbols in the input stream (as do some compilers). On the other hand, it may be desirable to clear the input stream of the remainder of the symbols of the rejected string.

4.7 Minimal FSRs

Each boxed segment of the program in Figure 4.18 represents one *state* in the corresponding FSR, $M_5{}^d$. In general, if we were to build such a parser for a regular language, we would like to know that our parser did not feature an unnecessarily large number of such segments.

```
V:   read(x)
     case x of
          eos:  goto NO
          'a'   goto X
          'b'   goto W
     endcase
```

```
X:   read(x)
     case x of
          eos:  goto NO
          'a'   goto Y
          'b'   goto W
     endcase
```

```
Y:   read(x)
     case x of
          eos:  goto NO
          'a'   goto H1
          'b'   goto W
     endcase
```

```
W:   read(x)
     case x of
          eos:  goto NO
          'a'   goto H2
          'b'   goto NLL
     endcase
```

```
H1:  read(x)
     case x of
          eos:  goto YES
          'a'   goto H1
          'b'   goto W
     endcase
```

```
H2:  read(x)
     case x of
          eos:  goto YES
          'a'   goto H2
          'b'   goto NLL
     endcase
```

```
NLL: read(x)
     case x of
          eos:  goto NO
          'a'   goto NLL
          'b'   goto NLL
     endcase
```

```
YES: write("string accepted")
exit.
NO:  write ("string rejected")
exit.
```
This part of the code is added to indicate success or failure.

Figure 4.18 The correspondence between a deterministic FSR ($M_5{}^d$ of Figure 4.15) and a decision program for the equivalent regular language.

Now, our program (Figure 4.18) *is* structured, notwithstanding misgivings often expressed about the use of unconditional jumps. If an equally structured program, consisting of fewer such segments, but doing the same job, could be derived, then we would prefer, for obvious and sound reasons, to use that one. In terms of a given FSR, the question is this: can we find an equivalent machine that features fewer states? We will go beyond this however: we will discover a method for producing a machine, from a given FSR, which is deterministic and uses *the smallest number of states necessary for a deterministic FSR to recognise the same language as that recognised by the original machine.* We call such a machine the *minimal FSR.*

If we produce such a minimal FSR, we can be confident that the corresponding decision program uses the smallest number of segments possible to perform the parsing task. Of course, there may be other methods for reducing the amount of code, but these are not of interest here.

4.7.1 Constructing a Minimal FSR

To demonstrate our method for producing minimal FSRs, we will use an FSR derived from a new example grammar, G_6, with productions:

$$S \to aA \,|\, bB$$
$$A \to aA \,|\, bB \,|\, bD$$
$$B \to bC \,|\, bB \,|\, bD$$
$$C \to cC \,|\, c$$
$$D \to bC \,|\, bD.$$

From G_6, we derive the non-deterministic FSR, M_6, which is shown in Figure 4.19.

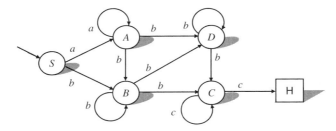

Figure 4.19 The non-deterministic FSR, M_6.

We practise our *subset algorithm* on the machine in Figure 4.19, to obtain the machine in Figure 4.20(a). For convenience, we rename the states of this deterministic machine to give us the machine in Figure 4.20(b) and which we call $M_6{}^d$.

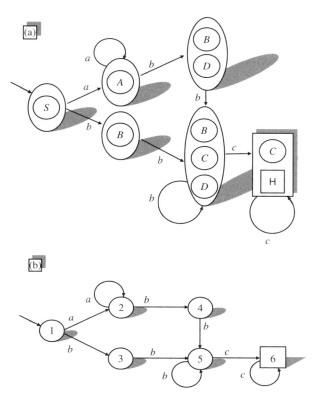

Figure 4.20 (a) $M_6{}^d$, the output of the subset algorithm for input M_6; and (b) the machine in (a) with renamed states.

For convenience, we omit the *null state* from the machines in this section.

Suppose we wished to produce a decision program for $M_6{}^d$, as we did for $M_5{}^d$ in the previous section. Our program would have seven segments of the type discussed earlier, as our machine $M_6{}^d$ has seven states including the null state. We will shortly see that we can produce an equivalent machine that has only five states, but first, we present, in Table 4.7, a simple procedure for creating the "reverse" machine, which will be used in our "minimisation" method.

Table 4.8 specifies an algorithm that uses our *reverse* (Table 4.7) and *subset* (Table 4.6) procedures to produce a *minimal* machine, M^{min}, from a deterministic FSR, M.

We can demonstrate the operation of the above method on our example machine $M_6{}^d$ from above (we use the one with the renamed states):

Table 4.7 Deriving M^{rev}, the "reversed" form of an FSR, M.

Step	Description
1	Let M^{rev} be an exact copy of M.
2	Objects of this form in M ... ... become objects of this form in M^{rev}

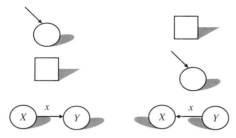

Note that all other parts of M^{rev} remain unchanged.

If L is the language recognised by M, then the language L^{rev}, recognised by M^{rev}, is L *with each and every sentence reversed.*

Table 4.8 "Minimising" an FSR, M.

Step	Description
1	*Reverse* M, using the method specified in Table 4.7.
2	Take the machine resulting from step 1, and apply the *subset algorithm* (Table 4.6) to create an equivalent deterministic machine.
3	*Reverse* the result of step 2.
4	Create an equivalent deterministic machine to the result of step 3.

The result of step 4, M^{min}, is the *minimal finite state recogniser* for the language recognised by M. There is no deterministic FSR with fewer states than M^{min} that could do the same job.

Figure 4.21 $M_6{}^d$, of Figure 4.20(b), reversed.

Step 1. Reverse $M_6{}^d$ (Figure 4.20(b)), giving the machine shown in Figure 4.21.

Step 2. Make the result of step 1 (Figure 4.21) deterministic. The result of step 2, after renaming the states, can be seen in Figure 4.22(b).

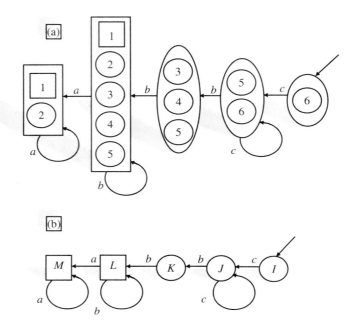

Figure 4.22 (a) The deterministic version of Figure 4.21; and (b) the machine in (a) with renamed states.

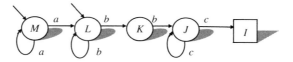

Figure 4.23 The reverse construction of Figure 4.22(b).

Step 3. Reverse the result of step 2. The reversed version of Figure 4.22(b) can be seen in Figure 4.23.

Note that the machine in Figure 4.23 has *two* start states; this is permitted in our rules for FSR behaviour and in rule 1 of the subset algorithm. However, this represents a non-deterministic situation, since such a machine effectively has a choice as to which state it starts in.

Now for the final step of the minimisation process:

Step 4. Make the result of step 3 (Figure 4.23) deterministic. We obtain the machine depicted in Figure 4.24(a).

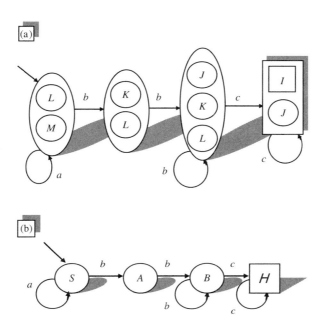

Figure 4.24 (a) The deterministic version of Figure 4.23; and (b) the machine in
(a) with renamed states, which is M_6^{min}, the minimal version of M_6.

The machine in Figure 4.24(b), resulting from renaming the states of the machine
in Figure 4.24(a), has been called M_6^{min}. It is the *minimal deterministic FSR* for
the language recognised by M_6^d.

We now see clearly that the language recognised by M_6^{min} (and, of course, M_6
and M_6^d), is:

$$\{a^i b^j c^k : i \geq 0, j \geq 2, k \geq 1\},$$

i.e. all sentences are of the form:

zero or more as, followed by two or more bs, followed by one or more cs.

With a little thought, we can easily justify that the machine is minimal. We must
have one state for the "zero or more as" part, a further *two* states for the "two or
more bs" part, and an extra state to ensure the "one or more cs" part.

4.7.2 Why Minimisation Works

The minimisation algorithm of Table 4.8 simply uses the *subset algorithm* twice,
first applying it to the reversed form of the original machine, and then to the

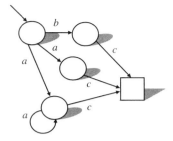

Figure 4.25 An FSR which will have a non-deterministic reverse construction.

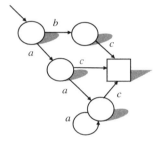

Figure 4.26 A deterministic version of Figure 4.25.

reversed form of the deterministic version of the original reversed form. Let us consider why this approach produces the correct result. To do this, we focus on the example FSR in Figure 4.25.

Making the machine in Figure 4.25 deterministic, by using our subset algorithm, we merge the two paths labelled a into one. We then obtain the machine in Figure 4.26.

Now, while the machine in Figure 4.26 is deterministic, it is not *minimal*. It seems that we ought to be able to deal with the fact that, regardless of whether the first symbol is an a or a b, the machine could get to its halt state given one c. However, as this is not a non-deterministic situation, the subset algorithm has ignored it.

Now, reversing the machine in Figure 4.26 yields the machine in Figure 4.27.

The "three cs situation" has become non-deterministic in Figure 4.27. The subset algorithm *will* now deal with this, resulting in the machine of Figure 4.28.

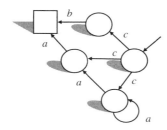

Figure 4.27 The reverse construction of Figure 4.26.

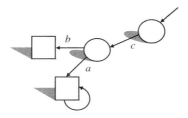

Figure 4.28 A deterministic version of Figure 4.27.

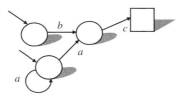

Figure 4.29 The reverse construction of Figure 4.28. (Note that this is also an
example of an FSR with two start states.)

While the machine in Figure 4.28 now consists of only four states and is
deterministic, it unfortunately recognises our original language in reverse, so we
must reverse it again, giving Figure 4.29.

The Figure 4.29 machine now recognises our original language, but is non-
deterministic (it has two start states *and* a state with two outgoing arcs labelled *a*).
So, we make it deterministic again, giving the machine of Figure 4.30.

The machine in Figure 4.30 is the minimal (deterministic) machine.

The algorithm thus removes non-determinism "in both directions" through
the machine. Reversing a machine creates a machine that recognises the

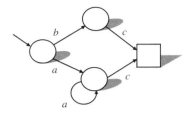

Figure 4.30 Deterministic version of Figure 4.29, the minimal version of
Figure 4.25.

original language with each and every string *reversed*, and the subset algorithm
always results in a machine that accepts exactly the same strings as the
machine to which it is applied. Thus, making the reversed machine deterministic and then reversing the result will not change the language recognised by the
machine.

There are other ways of creating a minimal FSR. One of the most popular is
based on *equivalence sets*. The beauty of the method we use here is that it is very
simple to specify and makes use of a method we defined earlier, i.e. the *subset
algorithm*. Thus, the minimisation method we have used here has a pleasing
modularity. It uses the *subset* and *reverse* methods as *subroutines* or *components*,
so to speak.

4.8 The General Equivalence of Regular Languages and FSRs

We saw above that any *regular grammar* can be converted into an equivalent
FSR. In fact, the converse is also true, i.e.,

> every FSR can be converted into a regular grammar.

We will consider an example showing the conversion of an FSR into regular
productions. The interesting thing about our example is that the particular
machine in question has transitions labelled ε. These are called *empty moves* or
ε *moves*, and enable an FSR to make a transition without reading an input
symbol. As we will see, they produce slightly different productions from the
ones we have seen up to now. We consider this observation in more detail after
working through the following example, which is based on the machine *M* shown
in Figure 4.31.

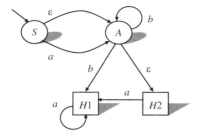

Figure 4.31 An FSR, M, with "empty moves" ("ε *moves*").

The grammar that we obtain from M, in Figure 4.31, which we will call G_m, is as follows:

$$S \rightarrow aA \,|\, A$$
$$A \rightarrow bA \,|\, bX \,|\, Y$$
$$X \rightarrow aX \,|\, \varepsilon$$
$$Y \rightarrow aX \,|\, \varepsilon.$$

Some issues are raised by the above construction:

- M can get to its halt state $H1$ without reading any input. Thus, ε is in the language accepted by M. As we would therefore expect, G_m generates ε $(S \rightarrow A \rightarrow Y \rightarrow \varepsilon)$.

- The *empty move* from S to A becomes a production $S \rightarrow A$. Productions of the form $x \rightarrow y$ where x and y are both non-terminals are called *unit productions*, and are actually prohibited by the definition of regular grammar productions with which we began this chapter. However, we will see in Chapter 5 that there is a method for removing these productions, leaving a grammar that *does* conform to our specification of regular grammar productions, and generates the same language as the original.

- M has arcs *leaving* halt states, and has more than one halt state. We saw above that the subset algorithm produces machines with these characteristics. In any case, there is nothing in the definition of the FSR that precludes multiple halt *or* start states. Where there is an arc leaving a halt state, such as the arc labelled a leaving $H2$, the machine could either halt or read an a. Halting in such a situation is exactly the same as an empty move, so in the grammar the $H2$ situation becomes $Y \rightarrow aX \,|\, \varepsilon$, where Y represents $H2$ and X represents $H1$.

I leave it to you to convince yourself that $L(G_m)$ is identical to the language accepted by M, and that the way the productions of G_m were obtained from M

could be applied to any FSR, no matter how complex, and no matter how many empty moves it has. In the above machine, the start state was conveniently labelled S. We can always label the states of an FSR how we please; a point made earlier in this chapter. Note that, for FSRs with more than one start state we can simply introduce a new *single* start state, with outgoing arcs labelled ε leaving it and directly entering each state that was a start state of the original machine (which we then no longer class as start states). Then the machine can be converted into productions, as was M, above.

The above should convince you that the following statement is true:

> for any FSR, M, there exists a regular grammar, G, which is such that $L(G)$, the language generated by G, is equal to the set of strings accepted by M.

Our earlier result that for every regular grammar there is an *equivalent* FSR means that if we come across a language that *cannot* be accepted by an FSR, we know that we are wasting our time trying to create a regular grammar to generate that language. As we will see in Chapter 6, there are properties of the FSR that we can exploit to establish that certain languages could not be accepted by the FSR. We now know that if we do this, we have also established that we could not define a regular grammar to generate that language. Conversely, if we can establish that there is no regular grammar to generate a given language, we know it is pointless attempting to create an FSR, or a parser of "FSR-level-power", like the decision program earlier in the chapter, to accept that language.

4.9 Observations on Regular Grammars and Languages

The properties of regular languages described in this chapter make them very useful in a computational sense. We have seen that simple *parsers* can be constructed for languages generated by a regular grammar. We have also seen how we can make our parsers extremely efficient, by ensuring that they are deterministic, reduced to the smallest possible form, and reflect a modular structure. We have also seen that there is a general *equivalence* between the FSRs and the regular languages.

Unfortunately, the simple structure of regular grammars means that they can only be used to represent simple constructs in programming languages, for example basic objects such as *identifiers* (variable names) in Pascal programs. In fact, the definition of Pascal "identifiers" given in context free form in Chapter 2 could be represented as a regular grammar. The regular languages can also be represented as symbolic statements known as *regular expressions*. These are not only a purely formal device, but are also implemented in text editors such as "vi" (Unix) and "emacs".

We will defer more detailed discussion of the limitations of regular grammars until Chapter 6. In that chapter we show that even a language as simple as $\{a^i b^i : i \geq 1\}$, generated by the two-production *context free* grammar $S \to aSb \mid ab$ (grammar G_3 of Chapters 2 and 3) is not regular. If such a language cannot be generated by a regular grammar, then a regular grammar is not likely to be capable of specifying syntactically valid Pascal programs, where, for example, a parser has to ensure that the number of *begin*s is the same as the number of *end*s. In fact, when it comes to programming languages, the *context free grammars* as a whole are much more important.[2] Context free grammars and context free languages are the subject of the next chapter.

EXERCISES

For exercises marked "†", solutions, partial solutions, or hints to get you started appear in "Solutions to Selected Exercises" at the end of the book.

4.1. Construct FSRs *equivalent* to the following grammars. For each FSR briefly justify whether or not the machine is *deterministic*. If the machine is *non-deterministic*, convert it into an equivalent deterministic machine.

(a) [†]

$$S \to aA$$
$$A \to aS \mid aB$$
$$B \to bC$$
$$C \to bD$$
$$D \to b \mid bB$$

Note: This is the grammar from Chapter 2, exercise 1(a).

---

[2] Bear in mind that the class of the context free grammars includes the class of the regular grammars.

(b)

$$S \rightarrow aA \,|\, aB \,|\, aC$$

$$A \rightarrow aA \,|\, aC \,|\, bC$$

$$B \rightarrow aB \,|\, cC \,|\, c$$

$$C \rightarrow cC \,|\, c$$

(c)   Grammar G_1 from Chapter 2, i.e.

$$S \rightarrow aS \,|\, bB$$

$$B \rightarrow bB \,|\, bC \,|\, cC$$

$$C \rightarrow cC \,|\, c$$

4.2. Given the FSR in Figure 4.32,

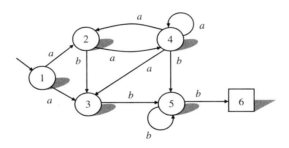

Figure 4.32 A finite state recogniser.

(a) Produce a *minimal* version.

(b) Construct an equivalent regular grammar.
      Hint: it is often more straightforward to use the original, non-deterministic machine (in this case, the one in Figure 4.32) to obtain the productions of the grammar, i.e., we write down the productions that would have given us the above machine according to our conversion rules.

4.3.[†] Design suitable data structures to represent FSRs. Design a program to apply operations such as the *subset method* and the *minimisation algorithm* to FSRs.

4.4 The *subset algorithm* is so called because each new state of M^d, the deterministic machine, produced contains a *subset* of the states of M. An *algorithm* is a procedure that always terminates for valid inputs. Consider why we can be sure that the subset procedure will terminate, given any non-deterministic machine as its input, and thus deserves to be called an algorithm.

Note: *this has something to do with the necessary limit on the number of new, composite states that can be created by the procedure. This point is explored in more detail in Chapter* 12.

5

Context Free Languages and Pushdown Recognisers

5.1 Overview

This chapter considers *context free grammars* (*CFG*s) and *context free languages* (*CFL*s).

First, we consider two methods for changing the productions of CFGs without changing the language they generate. These two methods are:

- removing the empty string (ε) from a CFG

- converting a CFG into *Chomsky Normal form*.

We find that, as for *regular languages*, there is an *abstract machine* for the CFLs called the *pushdown recogniser* (*PDR*), which is like a *finite state recogniser* equipped with a storage "device" called a *stack*.

We consider:

- how to create a *non-deterministic* PDR from a CFG

- the difference between *deterministic* and *non-deterministic* PDRs

- *deterministic* and *non-deterministic* CFLs

- the implications of the fact that not all CFLs are *deterministic*.

A.P. Parkes, *A Concise Introduction to Languages and Machines*,
DOI: 10.1007/978-1-84800-121-3_5, © Springer-Verlag London Limited 2008

5.2 Context Free Grammars and Context Free Languages

Recall from Chapter 2 that the context free grammars (type 2 in the *Chomsky hierarchy*) are *Phrase Structure Grammars* (*PSGs*) in which each and every production conforms to the following pattern:

$$x \to y, x \in N, y \in (N \cup T)^*$$ i.e. x is a single *non-terminal*, and y is an

arbitrary (possibly empty) mixture of

terminals and/or *non-terminals*.

So, the grammar G_2 of Chapter 2, i.e.,

$$S \to aB \mid bA \mid \varepsilon$$
$$A \to aS \mid bAA$$
$$B \to bS \mid aBB$$

is context free. If G is a CFG, then $L(G)$, the *language generated by the grammar G*, is a *context free language* (*CFL*).

We will consider a further example CFG, G_7:

$$S \to ABC$$
$$A \to aAb \mid B$$
$$B \to bB \mid \varepsilon$$
$$C \to ccC \mid B$$

$L(G_7) = \{a^h b^i c^{2j} b^k : h, j, k \geq 0, i \geq h\}$,   i.e., 0 or more *a*s followed by at least as

many *b*s, followed by an even number of

(or no) *c*s, followed by 0 or more *b*s,

as you may like to justify for yourself. You may arrive at a different, though equivalent, set definition; if so, check that it describes the same set as the one above.

5.3 Changing G without Changing $L(G)$

The purpose of this section is to show how one can modify the productions of a CFG in various ways without changing the language the CFG generates. There are more ways of modifying CFGs than are featured here, but the two methods below are relatively simple to apply. The results of this section serve also to simplify various proofs in Chapter 6.

5.3.1 The Empty String (ε)

Our example grammar G_7, above, features the *empty string* (ε) on the right-hand side of one of its productions. The empty string often causes problems in parsing and in terms of *empty moves* in finite state recognisers (FSRs), as was noted in Chapter 3. For parsing purposes, we would prefer the empty string not to feature in our grammars at all. Indeed, you may wonder whether the empty string actually features in any "real world" situations, or if it is included for mathematical reasons. Consider, for example string *concatenation*, where ε can be considered to play a similar part as 0 in addition, or 1 in multiplication.

The empty string has its uses, one of these being in the definition of programming languages. Pascal, for example, allows a <statement> to be empty (i.e., a statement can be an *empty string*). Such "empty" statements have their practical uses, such as when the programmer leaves empty "actions" in a *case* statement, or an empty *else* action in an *if* statement, to provide a placeholder for additional code to be entered later as needed. Note, however, that the empty string is not a *sentence* of the Pascal language. The Pascal syntax does not allow for an empty <program>, and thus, according to the definition of the language, an empty source code file should result in an error message from the Pascal compiler.

It turns out that for CFGs we can deal with the empty string in one of two ways. Firstly, for *any* phrase structure grammar (see Chapter 2), G, we can, by applying the rules we are about to see, remove ε altogether from the productions, except for one special case which we discuss later. We will need the terminology defined in Table 5.1 to discuss the empty string in CFGs. The table also features examples based on grammar G_7 from above.

Table 5.1 Terminology for discussing the empty string in grammars.

Term	Definition	Examples from G_7
ε *production*	Any production with ε on the right-hand side	$B \to \varepsilon$
ε-*generating non-terminal*	Any non-terminal, X, which is such that $$X \Longrightarrow^* \varepsilon$$ (i.e. any non-terminal from which we can generate the empty string)	S, A, B, and C since we can have: $S \Longrightarrow ABC$ and $A \Longrightarrow B \Longrightarrow \varepsilon$ and $C \Longrightarrow B \Longrightarrow \varepsilon$
ε-*free production*	Any production which does not have ε on the right-hand side	All productions except $B \to \varepsilon$

Table 5.2 How to remove ε from a CFG.

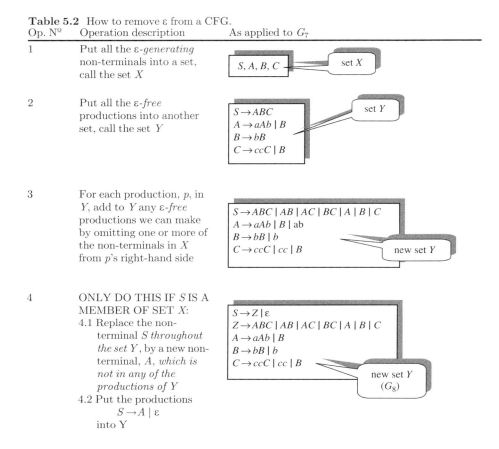

Op. Nº	Operation description	As applied to G_7
1	Put all the ε-*generating* non-terminals into a set, call the set X	
2	Put all the ε-*free* productions into another set, call the set Y	
3	For each production, p, in Y, add to Y any ε-*free* productions we can make by omitting one or more of the non-terminals in X from p's right-hand side	
4	ONLY DO THIS IF S IS A MEMBER OF SET X: 4.1 Replace the non-terminal S throughout the set Y, by a new non-terminal, A, *which is not in any of the productions of Y* 4.2 Put the productions $S \rightarrow A \mid ε$ into Y	

The method for removing ε *productions* from a grammar is described in Table 5.2. On the right of the table, we see the method applied to grammar G_7.

You should verify for yourself that the new grammar, which in Table 5.2 is called G_8, generates exactly the same set of strings as the grammar with which we started (i.e., G_7).

Referring to Table 5.2, at step 3, the productions added to the set Y were necessary to deal with cases where one or more non-terminals in the original productions could have been used to generate ε. For example, we have the production $S \rightarrow ABC$. Now in G_7, the B could have been used to generate ε, so we need to cater for this by introducing the production $S \rightarrow AC$. Similarly, both the A and the C could have been used to generate ε, so we need a production $S \rightarrow B$, and so on. Of course, we still need to include $S \rightarrow ABC$ itself, because, in general, the ε-generating non-terminals may generate other things apart from ε as they do in our example).

Note that step 4 of Table 5.2 is only applied if ε is in the language generated by the grammar (in formal parlance, if $ε ∈ L(G)$). This occurs only when S itself is an ε-generating non-terminal, as it is in the case of G_7. If S is an ε-generating non-terminal for a grammar G, then obviously ε is in $L(G)$. Conversely, if S is not such a non-terminal, then ε cannot be in the language. If we applied step 4 to a grammar for which ε is not in the corresponding language, the result would be a new grammar that did not generate the same language (it would generate the original set of strings along with ε, formally $L(G) ∪ \{ε\}$).

Alternatively, if we did not apply step 4 to a grammar in which S *is* an ε-generating non-terminal, then we would produce a grammar which generated all the strings in $L(G)$ except for the string ε (formally $L(G)-\{ε\}$). We must always take care to preserve exactly the set of strings generated by the original grammar.

If we were to apply the above rules to a *regular grammar*, the resulting grammar would also be regular. This concludes a discussion in Chapter 4 suggesting that the empty string in regular grammars need not be problematical.

For *parsing*, the grammars produced by the above rules are useful, since at most they contain one ε–production, and if they do it is $S →ε$. This means that when we parse a string x:

- if x is non-empty, our parser need no longer consider the empty string; it will not feature in any of the productions that could have been used in the derivation of x,

- if, on the other hand, $x = ε$, either our grammar has the one ε production $S → ε$, in which case we accept x, or the grammar has no ε productions, and we simply reject it.

For parsing purposes our amended grammars may be useful, but there may be reasons for leaving the empty string in a grammar showing us the structure of, say, a programming language. One of these reasons is readability. As we can see, G_8 consists of many more productions (16) than does the equivalent G_7 (7), and thus it could be argued that G_7 is more readily comprehensible to the reader. A format for grammars that is useful for a parser may not necessarily support one of the other purposes of grammars, i.e., to show *us* the structure of languages.

To complete this section, we will consider a further example of the removal of ε productions from a CFG. In this case ε is not in the language generated by the grammar. Only a sketch of the process will be given; you should be able to follow the procedure by referring to the rules in Table 5.2.

The example grammar, G_9, is as follows:

$$S → aCb$$

$$C → aSb \mid cC \mid ε.$$

In this case, the only ε-generating non-terminal is C. S cannot generate ε, as any string it generates must contain at least one a and one b. The new productions we would add are $S \rightarrow$ ab and $C \rightarrow$ c, both resulting from the omission of C. Step 4 of Table 5.2 should not be applied in this case, as S is not ε generating. The final amended productions would be:

$$S \rightarrow aCb \,|\, ab$$
$$C \rightarrow aSb \,|\, cC \,|\, c$$

A set definition of the language generated by this grammar, which, of course, is exactly the same as $L(G_9)$, is:

$$\{a^{2i+1}c^jb^{2i+1} : i, j \geq 0\} \quad \text{i.e., a non-zero odd number of } as, \text{followed by } 0$$
$$\text{or more } cs, \text{followed by the same number of } bs$$
$$\text{as there were } as.$$

5.3.2 Chomsky Normal Form

Once we have removed the empty string (ε) from a context free grammar, isolating it if necessary in one production, $S \rightarrow \varepsilon$, there are various methods for transforming the ε–*free* part of G while leaving $L(G)$ unchanged. One of the most important of these methods leaves the productions in a form we call *Chomsky Normal Form* (*CNF*). This section briefly describes the method for converting a CFG into CNF by presenting a worked example. The fact that we can assume that any CFG can be converted into CNF is useful in Chapter 6.

As an example grammar, we use part of G_8 (the ε-free version of G_7), and assume in this case that Z is the *start symbol*:

$$Z \rightarrow ABC \,|\, AB \,|\, AC \,|\, BC \,|\, A \,|\, B \,|\, C$$
$$A \rightarrow aAb \,|\, B$$
$$B \rightarrow bB \,|\, b$$
$$C \rightarrow ccC \,|\, cc \,|\, B.$$

The method to "Chomskyfy" a grammar is as follows. There are three steps, as now explained.

Step 1: Unit productions.

Unit productions were briefly introduced in Chapter 4. A unit production is a production of the form: $x \rightarrow y$, where x and y are single non-terminals. Thus, G_8 has the following unit productions:

$$Z \rightarrow A \mid B \mid C$$
$$A \rightarrow B$$
$$C \rightarrow B.$$

To help us deal with the unit productions of $G8$, we draw the structure shown in Figure 5.1. This represents all derivations that begin with the application of a unit production and end with a string that does not consist of a single non-terminal (each path stops the first time such a string is encountered).

The structure in Figure 5.1 is not a *tree*, but a *graph* (in fact, like an FSR, it is a directed graph, or *digraph*). We can call the structure a *unit production graph*. It could be drawn as a tree if we repeated some *nodes* at lower levels (in the graph in Figure 5.1 we would repeat B beneath A and beneath C). However, such a representation would not be so useful in the procedure for creating new non-unit productions described below.

Some unit production graphs we draw may lead to even more circular (*"cyclical"*), situations, as in the example in Table 5.3.

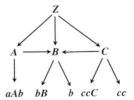

Figure 5.1 The *unit production graph* for grammar G_8.

Table 5.3 Example productions and a corresponding unit production graph.

Example productions	Resulting graph
$S \rightarrow aS \mid B$	
$B \rightarrow bS \mid A \mid C$	
$A \rightarrow a \mid C$	
$C \rightarrow bA \mid S \mid B$	

We now continue with our initial example. From the unit production graph for G_8 (Figure 5.1) we obtain:

- every possible non-unit production we can obtain by following a directed path from every non-terminal to every node that does not consist of a single non-terminal, where each production we make has a left-hand side that is the non-terminal we started at, and a right-hand side that is the string where we stopped.

So, from our unit production graph for G_8 in Figure 5.1, we obtain the following productions (those crossed out are already in G_8, so we need not add them):

$$Z \rightarrow bB\,|\,b\,|\,aAb\,|\,ccC\,|\,cc \qquad \{\text{by following all paths from } Z\}$$

$$A \rightarrow aAb\,|\,bB\,|\,b \qquad\qquad \{\text{by following all paths from } A\}$$

$$B \rightarrow bB\,|\,b \qquad\qquad\qquad \{\ldots\text{from } B\}$$

$$C \rightarrow ccC\,|\,cc\,|\,bB\,|\,b \qquad\quad \{\ldots\text{and from } C\}.$$

We remove all of the unit productions from the grammar and add the new productions to it, giving $G_{8(a)}$ (new parts are underlined):

$$Z \rightarrow ABC\,|\,AB\,|\,AC\,|\,BC\,|\,\underline{bB}\,|\,\underline{b}\,|\,\underline{aAb}\,|\,\underline{ccC}\,|\,\underline{cc}$$

$$A \rightarrow aAb\,|\,\underline{bB}\,|\,\underline{b}$$

$$B \rightarrow bB\,|\,b$$

$$C \rightarrow ccC\,|\,cc\,|\,\underline{bB}\,|\,\underline{b}.$$

You should perhaps take a moment to convince yourself that the terminal strings we can generate from $G_{8(a)}$ are exactly the same as those we can generate starting from Z in G_8 (remember that Z is the start symbol, in this case). Observe, for example, that from the initial grammar, we could perform the derivation $Z \Longrightarrow A \Longrightarrow B \Longrightarrow C \Longrightarrow ccC$. Using $G_{8(a)}$, this derivation is simply $Z \Longrightarrow ccC \Longrightarrow cccc$.

In Chapter 4, with reference to the conversion of an arbitrary *finite state recogniser* into *regular grammar* productions, we noted that *empty moves* in an FSR (transitions between states that require no input to be read, and are thus labelled with ε) correspond to unit productions. You can now see that if we applied step 1 of "Chomskyfication" to the ε-free part of a regular grammar containing unit productions, we would obtain a grammar that was also regular.

Step 1 is now complete, as unit productions have been removed. We turn now to step 2.

Step 2: Productions with terminals and non-terminals on the right-hand side, or with more than one terminal on the right-hand side.

Our grammar $G_{8(a)}$ now contains no unit productions. We now focus on the productions in $G_{8(a)}$ described in the title to this step. These productions we call *secondary productions*. In $G_{8(a)}$ they are:

$$Z \rightarrow bB \,|\, aAb \,|\, ccC \,|\, cc$$
$$A \rightarrow aAb \,|\, bB$$
$$B \rightarrow bB$$
$$C \rightarrow ccC \,|\, cc \,|\, bB.$$

To these productions we do the following:

- replace each terminal, t, by a new non-terminal, N, and introduce a new production $N \rightarrow t$, in each case.

So, for example, from $A \rightarrow aAb \,|\, bB$, we get:

$$A \rightarrow JAK \,|\, KB$$
$$J \rightarrow a$$
$$K \rightarrow b.$$

Now we can use J and K to represent a and b, respectively, throughout our step 2 productions, adding all our new productions to $G_{8(a)}$ in place of the original step 2 productions. Then we replace c by L in all the appropriate productions, and introduce a new production $L \rightarrow c$. The result of this we will label $G_{8(b)}$:

$$Z \rightarrow ABC \,|\, AB \,|\, AC \,|\, BC \,|\, b \,|\, \underline{KB} \,|\, \underline{JAK} \,|\, \underline{LLC} \,|\, \underline{LL}$$
$$A \rightarrow \underline{JAK} \,|\, \underline{KB} \,|\, b$$
$$B \rightarrow \underline{KB} \,|\, b$$
$$C \rightarrow \underline{LLC} \,|\, \underline{LL} \,|\, \underline{KB} \,|\, b$$
$$\underline{J \rightarrow a}$$
$$\underline{K \rightarrow b}$$
$$\underline{L \rightarrow c.}$$

The new productions are again underlined. You can verify that what we have done has not changed the terminal strings that can be generated from Z.

We are now ready for the final step.

Step 3: Productions with more than two non-terminals on the right-hand side.

After steps 1 and 2, all our productions will be of the form:

$$x \to y \text{ or } x \to z \quad \text{where } y \text{ consists of two or more } \textit{non-terminals}$$

$$\text{and } z \text{ is a single } \textit{terminal}.$$

As the step 3 heading suggests, we now focus on the following productions of $G_{8(b)}$:

$$Z \to ABC \,|\, JAK \,|\, LLC$$

$$A \to JAK$$

$$C \to LLC.$$

As an example consider $Z \to ABC$. It will not affect things if we introduce a new non-terminal, say P, and replace $Z \to ABC$ by:

$$Z \to AP$$

$$P \to BC,$$

the $G_{8(b)}$ derivation $Z \implies ABC$ would then involve doing: $Z \implies AP \implies ABC$. As P is not found anywhere else in the grammar, this will not change any of the terminal strings that can be derived. We treat the rest of our step 3 productions analogously, again replacing the old productions by our new ones, and using a new non-terminal in each case. This gives us the grammar $G_{8(c)}$:

$$Z \to \underline{AP} \,|\, AB \,|\, AC \,|\, BC \,|\, b \,|\, KB \,|\, \underline{JQ} \,|\, \underline{LR} \,|\, LL$$

$$\underline{P \to BC}$$

$$\underline{Q \to AK}$$

$$\underline{R \to LC}$$

$$A \to \underline{JT} \,|\, KB \,|\, b$$

$$\underline{T \to AK}$$

$$B \to KB \,|\, b$$

$$C \to \underline{LU} \,|\, LL \,|\, KB \,|\, b$$

$$\underline{U \to LC}$$

$$J \to a$$

$$K \to b$$

$$L \to c.$$

(Changed and new productions have once again been indicated by underlining.)

The resulting grammar, $G_{8(c)}$, is in *Chomsky normal form.*

It is important to appreciate that, unlike in G_8, step 3 productions may include right-hand sides of *more than three* non-terminals. For example, we may have a production such as

$$E \rightarrow DEFGH.$$

In such cases, we split the production up into productions as follows:

$E \rightarrow DW$ where W, X and Y are new non-terminals, not found

$W \rightarrow EX$ anywhere else in the grammar.

$X \rightarrow FY$

$Y \rightarrow GH$

A derivation originally expressed as $E \Longrightarrow DEFGH$ would now involve the following:

$$E \Longrightarrow DW \Longrightarrow DEX \Longrightarrow DEFY \Longrightarrow DEFGH.$$

From this example, you should be able to induce the general rule for creating several "two-non-terminals-on-the-right-hand-side" productions from one "more-than-two-non-terminals-on-the-right-hand-side" production.

If we apply steps 1, 2 and 3 to any ε-free CFG, we obtain a CFG where all productions are of the form

$x \rightarrow y$ or $x \rightarrow uv$, where y is a single *terminal*, and x, u and v are single non-terminals.

A CFG like this is said to be in *Chomsky normal form (CNF)*.

It should therefore be clear that any ε-*free* CFG can be converted into CNF.

Once the ε-free part of the grammar has been converted into CNF, we need to include the two productions that resulted from step 4 of the method for removing the empty string (Table 5.2), *but only if step 4 was carried out for the grammar.* In the case of G_8, step 4 *was* carried out. To arrive at a grammar equivalent to G_8 we need to add the productions $S \rightarrow$ ε and $S \rightarrow Z$ to $G_{8(c)}$, and once again designate S as the start symbol.

It is very useful to know that any CFG can be converted into CNF. For example, we will use this "CNF assumption" to facilitate an important proof in the next chapter. In a CFG in CNF, all *reductions* will be considerably simplified, as the *right-hand sides* of productions consist of single terminals or two non-terminals.

A *parser* using such a grammar need only examine two adjacent non-terminals or a single terminal symbol in the *sentential form* at a time. This simplifies things considerably, since a *non-CNF* CFG could have a large number of terminals and/or non-terminals on the right-hand side of some of its productions.

A final point is that if the CFG is in CNF, any *parse tree* will be such that every non-terminal node expands into either two non-terminal nodes or a single terminal node. Derivations based on CNF grammars are therefore represented by *binary* trees. A binary tree is a tree in which no node has more than two outgoing arcs. A parser written in Pascal could use a three-field *record* to hold details of each non-terminal node that expands into two non-terminals (one field for the node's label, and two pointers to the two non-terminals into which the non-terminal expands). A single type of record structure would therefore be sufficient to represent all of the non-terminal nodes of the parse tree.

5.4 Pushdown Recognisers

In Chapter 4, we saw that from any *regular grammar* we can build an *abstract machine* called a *finite state recogniser*. The machine is such that it recognises every string (and no others), that is generated by the corresponding grammar. Moreover, for any finite state recogniser we invent we can produce an equivalent regular grammar. It was also said (though it will not be *proved* until Chapter 6) that there is at least one language, $\{a^i b^i : i \geq 1\}$, that cannot be recognised by any finite state recogniser (and therefore cannot be generated by a regular grammar), which nevertheless we have seen to be context free.

In terms of the processing of the *context free* languages, the problem with the finite state recogniser is that it has no memory device: it has no way of "remembering" details about the part of the string it has processed at a given moment in time. This observation is crucial in several results considered in this book. The FSR's only memory is in terms of the state it is in, which provides limited information. For example, if a finite state recogniser wishes to "count" n symbols, it has to have n states with which to do it. This immediately makes it unsuitable even for a simple context free language, such as $\{a^i b^i : i \geq 1\}$, that requires a recogniser to be able to "count" a number of *a*s that cannot be predicted in advance, so that it can make sure that an equal number of *b*s follows.

It turns out that with the addition of a rudimentary memory device called a *stack*, the finite state recogniser can do the job for context free languages that the finite state recogniser can do for regular languages. We call such an augmented finite state recogniser a *pushdown recogniser* (*PDR*). Before we examine these machines in detail, we need to be clear about what is meant by a *stack*.

5.4.1 The Stack

A stack is simply a linear data structure that stores a sequence of objects. In terms of this chapter, these "objects" are symbols from a formal alphabet. The stack can only be accessed by putting an object on, or removing an object from, one end of the stack that is usually referred to as the *top* of the stack. The stack has a special marker ($\perp$), called the *stack bottom marker* which is always at the *bottom* of the stack. We use a diagrammatic representation of a stack as shown in Figure 5.2. Formally, our stacks are really strings that have the symbol $\perp$ at the rightmost end, and to which we can add or remove symbols from the leftmost end.

Figure 5.3 specifies the operations *PUSH* and *POP* that are needed by our machines.

However, note that we cannot *POP* anything from an empty stack. We have defined a *destructive POP*, in that its result is the top element of the stack, but its *side effect* is to remove that element from the stack. Finally, observe from Figure 5.4 that pushing the *empty string* onto the stack leaves the stack unchanged.

The stacks in this chapter are used to store strings of terminals and non-terminals, such that each symbol in the string constitutes one element of the stack. For convenience, we allow *PUSH*(*abc*), for example, as an abbreviation of *PUSH*(*c*) followed by *PUSH*(*b*) followed by *PUSH*(*a*). Given that, as stated above, our stacks are really strings, we could have defined *POP* and *PUSH* as formal string functions.

Stacks are important data structures and have many applications throughout computer science. A related data structure you may encounter (though we have no cause to use it in this book) is called a *queue*. A queue is like a stack, except that, as its name suggests, items are added to one end, but taken from the other.

5.4.2 Constructing a Non-deterministic PDR

The rules we introduce in this section (in Table 5.4) will produce a *non-deterministic PDR* (*NPDR*) from the productions of any context free grammar. As for finite state

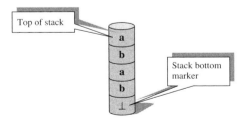

Figure 5.2 A *stack*. Symbols can only be added to, and removed from, the *top*.

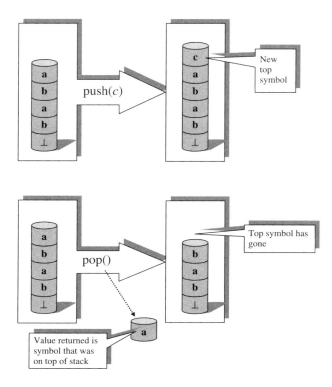

Figure 5.3 "push" and "pop" applied to the stack of Figure 5.2.

recognisers, the PDR has a pictorial representation. However, due to the additional complexity of the PDR resulting from the addition of the stack, the pictorial form does not provide such a succinct description of the language *accepted* by a PDR as it does in the case of the FSR. The rules in Table 5.4 convert any CFG, G, into a three-state NPDR.

The *acceptance conditions* for the NPDR, given some input string, x, are specified in Table 5.5.

As for FSRs (Chapter 4), we assert that the set of all *acceptable* strings of an NPDR derived from a grammar, G, is exactly the same as $L(G)$. We are not going to prove this, however, because as you will be able to see from the examples we consider next, it is reasonably obvious that any PDR constructed by using rules in Table 5.4 will be capable of reproducing any derivation that the grammar could have produced. Moreover, the PDR will reach a halt state only if its input string is a sentence that the grammar could derive. However, if you want to try to *prove* this, it makes a useful exercise.

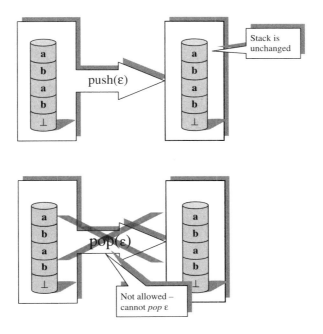

Figure 5.4 "push" and "pop" and the empty string (ε).

5.4.3 Example NPDRs, M_3 and M_{10}

Let us now produce an NPDR from the grammar G_3 that we encountered earlier, i.e.,

$$S \rightarrow aSb \mid ab.$$

Figure 5.5 shows the transitions produced by each rule (the rule number applied is in the upper right-hand corner of each box).

Figure 5.6 shows the connected machine (we have simply joined the appropriate circles together).

As in the previous chapter, we show the correspondence between the grammar and the machine by using the same subscript number in the name of both. We now investigate the operation of such a machine, by presenting M_3 with the input string a^3b^3. G_3 can generate this string, so M_3 should accept it. The operation of M_3 on the example string is shown in Table 5.6. Table 5.6(a) shows M_3's initial configuration, while Tables 5.6(b) and 5.6(c) show the machine's behaviour, eventually accepting the string in Table 5.6(c).

Notice in Tables 5.6(b) and 5.6(c) how either of the transitions derived from the productions $S \rightarrow aSb$ and $S \rightarrow ab$ can be applied any time we have S on the top

Table 5.4 How to create an NPDR from a CFG.

Step	Actions	Comments
1	We always begin with where S is the start symbol of the grammar.	Note that state 1 is the *start state*. The meaning of the labels on the *arcs* can be appreciated from the following: which means: if in state A, about to read a, and there is a b *on top of the stack*, a transition can be made to state B, with the symbol b being popped off the stack, and then the string x being pushed onto the stack. When *strings* are $PUSH$ed, for example, $PUSH(efg)$ is interpreted as $PUSH(g)$ then $PUSH(f)$ then $PUSH(e)$. So, the *transition* introduced in step 1 means: "when in state 1, *without reading any input*, and POPping the stack bottom marker off the stack, $PUSH$ the stack bottom marker back on the stack, then push S onto the stack, then go to state 2".
2	(a) For each production of G, $x \rightarrow y$, introduce a transition:	So, in state 2, we will have a situation where the left-hand side (i.e. non-terminal symbol) of any production of G that is at the top of the stack can be replaced by the right-hand side of that production. Again, all of these transitions are made without affecting the input string. As you may be able to see, the machine will operate by modelling derivations that could be carried out using the grammar.
	(b) For each terminal symbol, t, found in the grammar, G, introduce a transition:	Still in state 2, we now will have a situation where if a given terminal symbol is next in the input, and also on top of the stack, we can read the symbol from the input, and remove the one from the top of the stack (i.e. the effect of POPping the symbol off and then $PUSH$ing ε). These transitions allow us to read input symbols only if they are in the order that would have been produced by derivations made by the grammar (ensured by the transitions introduced by rule 2a).
3	Introduce a transition:	In state 2, we can move to the *halt state* (again represented as a box), only if the stack is empty. We make this move without reading any input, and we ensure that the stack bottom marker is replaced on the stack.

Table 5.5 The acceptance conditions for the NPDR.
The machine begins in the *start state*, with the stack containing only the *stack bottom marker*, $\perp$.

If there is a sequence of transitions by which the machine can reach its *halt state*, H, and on doing so, the input is *exhausted* and the stack contains only the stack bottom marker, $\perp$, then the string x is said to be *acceptable*.

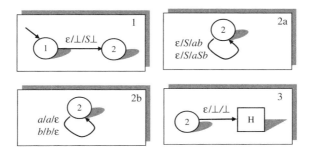

Figure 5.5 Creating a non-deterministic *pushdown recogniser* (*PDR*) for grammar G_3.

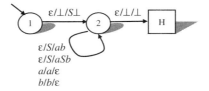

Figure 5.6 The NPDR, M_3.

of the stack. There are in fact (infinitely) many sequences of transitions that can be followed, but most of them will lead to a non-acceptance state.

We complete this section by considering a further example, this time a grammar that is equivalent to grammar G_2, of Chapters 2 and 3. G_2 is:

$$S \rightarrow aB \,|\, bA \,|\, \varepsilon$$

$$A \rightarrow aS \,|\, bAA$$

$$B \rightarrow bS \,|\, aBB.$$

However, our example, G_{10}, below has been constructed by applying our rules for removing the empty string (Table 5.2) to G_2. You should attempt this construction as an exercise.

Table 5.6(a) An initial configuration of the NPDR, M_3.

Initial state	Initial stack	Input

Table 5.6(b) Trace of M_3's behaviour (part 1), starting from the intial configuration in Table 5.6(a).

Transition	Resulting stack	Input
1 →(ε/⊥/S⊥) 2	S ⊥	$a\,a\,a\,b\,b\,b$
ε/S/aSb (2 loop)	a S b ⊥	$a\,a\,a\,b\,b\,b$
a/a/ε (2 loop)	S b ⊥	$a\,a\,a\,b\,b\,b$
ε/S/aSb (2 loop)	a S b b ⊥	$a\,a\,a\,b\,b\,b$
a/a/ε (2 loop)	S b b ⊥	$a\,a\,a\,b\,b\,b$

Table 5.6(c) Trace of M_3's behaviour (part 2), continuing from Table 5.6(b)

Transition	Resulting stack	Input

Input *exhausted*, in halt state, therefore input string *accepted*.

G_{10}'s productions are as follows:

$$S \rightarrow \varepsilon \mid X$$
$$X \rightarrow aB \mid bA$$
$$A \rightarrow aX \mid a \mid bAA$$
$$B \rightarrow bX \mid b \mid aBB.$$

We now apply the rules for creating the *equivalent* NPDR, resulting in the machine M_{10}, shown in Figure 5.7.

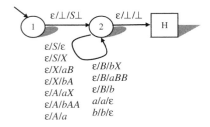

Figure 5.7 The NPDR, M_{10}.

Note that, unlike M_3, M_{10} will accept the empty string (since it allows S at the top of the stack to be replaced by ε), as indeed it should, as $\varepsilon \in L(G_{10})$ – it is one of the sentences that can be derived by G_{10}. However, $\varepsilon \notin L\ (G_3)$, i.e. the empty string is not one of the strings that can be derived from the start symbol of G_3.

I now leave it to you to follow M_{10}'s behaviour on some example strings. You are reminded that $L(G_{10})$ is the set of all strings of as and bs in any order, but in which the number of as is equal to the number of bs.

5.5 Deterministic Pushdown Recognisers

For the FSRs, *non-determinism* is not a critical problem. We discovered in Chapter 4 that for any non-deterministic FSR we could construct an equivalent, *deterministic FSR*. The situation is different for PDRs, as we will see. First, let us clarify how non-determinism manifests itself in PDRs. In PDRs, two things determine the next move that can be made from a given state:

1. the next symbol to be read from the input,

and

2. the symbol currently on top of the stack.

If, given a particular instance of 1 and 2, the machine has *transitions* that would allow it to:

A. make a choice as to which state to move to,

or

B. make a choice as to which string to push onto the stack

(or both), then the PDR is non-deterministic.

On the other hand, a *deterministic pushdown recogniser (DPDR)*, is such that

- in any state, given a particular input symbol and particular stack top symbol, the machine has only *one* applicable transition.

The question we consider here is: *for any NPDR, can we always construct an equivalent DPDR?* We saw in Chapter 4 that the answer to the corresponding question for FSRs was "yes". For PDRs however, the answer is more like "in general no, but in many cases, yes". We shall see that the "many cases" includes all of the regular languages and some of the *proper* CFLs (i.e. languages that are context free *but not regular*). However, we shall also see that there are some languages that are certainly context free, but *cannot* be recognised by a DPDR.

5.5.1 $M_3{}^d$, a Deterministic Version of M_3

First, we consider context free languages that *can* be recognised by a DPDR. The language:

$$A = \{a^i b^i : i \geq 1\},$$

is not regular, as was stated earlier and will be proved in Chapter 6. However, A is certainly context free, since it is generated by the CFG, G_3. Above, we constructed an NPDR (M_3) for A, from the productions of G_3 (see Figure 5.6). However, we can see that any sentence in the language A consists of a number of as followed by the same number of bs. A deterministic method for accepting strings of this form might be:

- PUSH each a encountered in the input string onto the *stack* until a b is encountered, then POP an a off the stack for each b in the input string.

The DPDR $M_3{}^d$, shown in Figure 5.8, uses the above method.

In Table 5.6 we traced the operation of the NPDR M^3 for the input a^3b^3. We now trace the operation of $M_3{}^d$ for the same input. Table 5.7(a) shows the initial configuration of $M_3{}^d$, Table 5.7(b) shows the behaviour of the machine until it accepts the string.

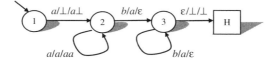

Figure 5.8 The DPDR, $M_3{}^d$.

Table 5.7(a) An initial configuration of the DPDR, $M_3{}^d$.

Initial state	Initial stack	Input

Table 5.7(b) Trace of $M_3{}^{d}$'s behaviour, starting from the initial configuration in Table 5.7(a).

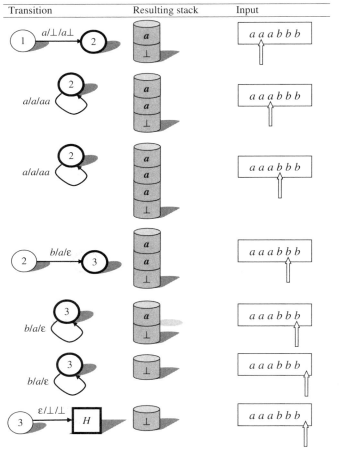

Transition	Resulting stack	Input

Input *exhausted*, in halt state, stack empty, therefore string *accepted*.

5.5.2 More Deterministic PDRs

Two more example DPDRs follow (Figures 5.9 and 5.10). It is suggested that you trace their behaviour for various valid and invalid input strings.

First, Figure 5.9 shows a DPDR for the language $L(G_{10})$, the set of all strings of as and bs in which the number of as is equal to the number of bs. This is the same language as $L(G_2)$, as we obtained the productions for G_{10} by applying our procedure for removing ε productions to G_2. Let us consider $L(G_{10})$ without the empty string (using the *set difference* operator, $L(G_{10})- \{\varepsilon\}$), which is obviously the set of all *non-empty* strings of as and bs in which the number of as is equal to the number of bs.

The DPDR shown in Figure 5.10 *recognises* the language $L(G_{10})$, which is:

$$\{a^{2i+1}c^jb^{2i+1} : i,j \geq 0\} \quad \text{i.e., a non-zero odd number of } a\text{s, followed by } 0$$
$$\text{or more } c\text{s, followed by the same number of } b\text{s}$$
$$\text{as there were } a\text{s.}$$

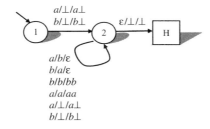

Figure 5.9 A DPDR for the language $L(G_{10}) - \{\varepsilon\}$, i.e., the language recognised by M_{10} but without the empty string.

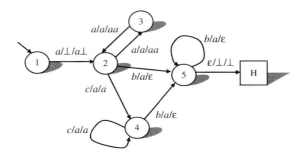

Figure 5.10 A DPDR for the language $L(G_9)$.

Clearly, then, there are CFLs that can be recognised by DPDRs. Here, we have encountered but a few. However, there are, as you can probably imagine, an infinite number of them. If a CFL can be recognised by a DPDR, we call it a *deterministic context free language*.

5.6 Deterministic and Non-deterministic Context Free Languages

In this section, we discover that there are context free languages that cannot be recognised by a deterministic PDR. There is no general equivalence between non-deterministic and deterministic PDRs, as there is for FSRs. We begin our investigation with regular languages, which we find to be wholly contained within the deterministic CFLs.

5.6.1   Every Regular Language is a Deterministic CFL

Any *regular language* is generated by a regular grammar without ε *productions* (apart from the one εproduction $S \rightarrow \varepsilon$, if it is necessary). Such a grammar can be represented as a *non-deterministic FSR*, which can, if necessary, be represented as a *deterministic FSR* (*DFSR*).

Now, any DFSR can be represented as a DPDR that effectively does not use its stack. We will use a single example that suggests a method by which any DFSR could be converted into an equivalent DPDR. The example deterministic FSR is $M_5{}^d$, from Chapter 4, shown in Figure 5.11.

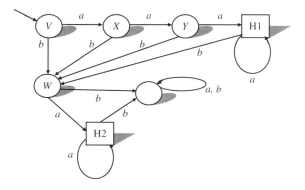

Figure 5.11 The DFSR, $M_5{}^d$, from Chapter 4.

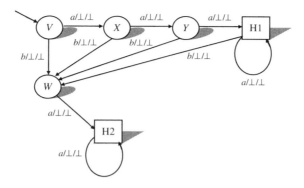

Figure 5.12 A DPDR equivalent to the DFSR, M_5^d, of Figure 5.11.

This produces the DPDR of Figure 5.12.

The DPDR in Figure 5.12 has two *halt states* (there is nothing wrong with this). We have also ignored the *null state* of the DFSR. We could clearly amend any DFSR to produce an equivalent DPDR as we have done above.

However, for the DPDR, we do need to amend our definition of the *acceptance conditions* as they applied to the NPDR (Table 5.5), which we derived by using rules that ensure the machine has only one halt state (Table 5.4).

The *acceptance conditions* for the DPDR, for an input string x, are specified in Table 5.8.

You may notice that the condition "the stack contains *only the stack bottom marker* $\perp$", that applied to NPDRs (Table 5.5) does not appear in Table 5.8. An exercise asks you to justify that the language $\{a^i b^j : i \geq 1, \ j = i \ or \ j = 0\}$ can only be recognised by a *deterministic* PDR if that machine is not required to have an empty stack every time it accepts a string.

So far, we have established the existence of the *deterministic context free languages*. Moreover, we have also established that this class of languages includes the regular languages. We have yet to establish that there are some CFLs that are necessarily *non-deterministic*, in that they can be accepted by an NPDR but not a DPDR.

Table 5.8 The acceptance conditions for the DPDR.

The machine begins in the *start state*, with the stack containing only the *stack bottom marker*, $\perp$.

If there is a sequence of transitions by which the machine can reach one of its *halt states*, and on doing so, the input is *exhausted*, then the string x is said to be *acceptable*.

5.6.2   The Non-deterministic CFLs

Consider the following CFG, G_{11}:

$$S \rightarrow a \mid b \mid c \mid aSa \mid bSb \mid cSc.$$

$L(G_{11})$ is the set of all possible non-empty odd-length *palindromic* strings of as
and/or bs and/or cs. A *palindromic* string is the same as the reverse of itself. We
could use our procedure from Table 5.4 to produce an NPDR to accept $L(G_{11})$.
However, could we design a DPDR to do the same? We cannot, as we now
establish by intuitive argument. General palindromic strings are of the form
specified in Figure 5.13.

The central box of Figure 5.13 could be empty if even-length palindromic
strings were allowed. Note that G_{11} does not generate even-length palindromes,
though it could be easily amended to do so, as you might like to establish for
yourself. Figure 5.14 shows how an example sentence from $L(G_{11})$ is partitioned
according to Figure 5.13.

Let us hypothesise that a DPDR could accept $L(G_{11})$. It would have to read
the input string from left to right, as that is how it works. An obvious deterministic way to do the task on that basis is as follows:

1. Read, and push onto the stack, each input symbol encountered until the
   "centre" of the string is reached.

2. Read the rest of the string, comparing each symbol to the next symbol on the
   stack. If the string is valid, the string on the stack following (1) will be the part
   of the string after the centre *in the correct order*.

The problem with the above is in the statement '*until the "centre" of the string is
reached*', in step (1). How can the DPDR detect the centre of the string? Of
course, it cannot. If there is a symbol in the centre it will be one of the symbols — a,
b, or c — that can feature *anywhere* in the input string. The machine must push
every input symbol onto the stack, on the assumption that it has not yet reached
the centre of the input string. This means that the machine must be allowed to
backtrack, once it reaches the end of the input, to a situation where it can see if it is

Figure 5.13 A specification of arbitrary palindromic strings.

Figure 5.14 A palindrome, partitioned according to Figure 5.13.

possible to empty its stack by comparing the stack contents with the remainder of the input. By similar argument, any such language of arbitrary palindromic strings defined with respect to an alphabet of two or more symbols is necessarily non-deterministic, while nevertheless being context free.

We have argued that the problem *vis-à-vis* palindromic strings lies in the PDR's inability to detect the "centre" of the input. If a language of palindromic strings is such that every string is of the form shown in Figure 5.15, then it can be accepted by a DPDR, and so is deterministic.

As an example of a deterministic palindromic language, consider $L(G_{12})$, where G_{12} (a slightly amended G_{11}) is as follows:

$$S \rightarrow d \,|\, aSa \,|\, bSb \,|\, cSc.$$

$L(G_{12})$ is the set of all palindromic strings of the form described in Figure 5.16.

$L(G_{12})$ is accepted by the DPDR M_{12} depicted in Figure 5.17.

Figure 5.15 Palindromes with a distinct central "marker".

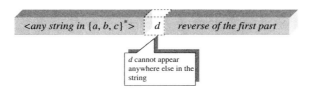

Figure 5.16 Sentences of $L(G_{12})$.

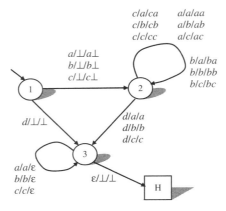

Figure 5.17 The DPDR, M_{12}, which deterministically accepts a language of palindromic strings that feature a central "marker".

5.6.3   A Refinement to the Chomsky Hierarchy in the Case of CFLs

The existence of *properly non-deterministic CFLs* suggests a slight refinement to the Chomsky hierarchy with respect to the context free and regular languages, as represented in Figure 5.18.

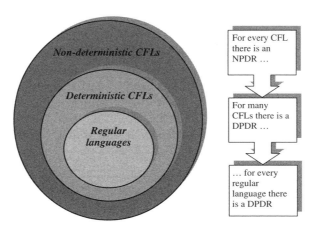

Figure 5.18 The context free languages: non-deterministic, deterministic and regular.

5.7 The Equivalence of CFGs and PDRs

The situation for FSRs and regular languages, i.e., that any regular grammar is represented by an *equivalent* FSR, and every FSR is represented by an equivalent regular grammar, is mirrored in the relationship between PDRs and CFGs. In this chapter, we have seen that for any CFG we can construct an equivalent NPDR. It turns out that for any NPDR we can construct an equivalent CFG. This will not be proved here, as the proof is rather complex, and it was considered more important to investigate the situation pertaining to the PDR as the *recogniser* for the CFLs, rather than the converse result.

One way of showing that every NPDR has an equivalent CFG is to show that from any NPDR we can create CFG productions that generate exactly the set of strings that enable the NPDR to reach its halt state, leaving its stack empty. The details are not important, but the result is powerful, as it shows us that any language that cannot be generated by a CFG (we see two such languages in the next chapter) needs a machine that is more powerful than a PDR to recognise it.

Consider the "multiplication language",

$$\{a^i b^j c^{i \times j} : i, j \geq 1\},$$

introduced in Chapter 2. This is shown to be not a CFL at the end of Chapter 6, a result that means that the language cannot be recognised by a PDR. This leads us to speculate that we need a more powerful machine than the PDR to model many computational tasks. We return to this theme in the second part of the book, when we consider abstract machines as *computers*, rather than language recognisers, and we see the close relationship between *computation* and language recognition.

5.8 Observations on Context Free Grammars and Languages

The lack of general equivalence between deterministic and non-deterministic context free languages is problematic for the design of programming languages. Efficient parsers can be designed from *deterministic* PDRs. As argued in Chapter 4, non-determinism is undesirable, as it is computationally expensive. Moreover, *unnecessary* non-determinism is simply wasteful in terms of resources. It turns out that any deterministic context free language can be parsed by a pushdown machine that never has to look ahead more than one symbol in the input to determine the choice of the next transition.

Consider, for example, sentences of the language $\{a^i b^i : i \geq 1\}$. If we look at the next symbol in the input, it will either be an a or a b, and, like our deterministic machine M_3, above, we know exactly what to do in either case. Now consider strings of arbitrary *palindromes*, with no distinct central "marker" symbol. In this case, for any machine that could look ahead a fixed number of symbols, we could simply present that machine with a palindromic string long enough to ensure that the machine could not tell when it moved from the first to the second "half" of the input. Languages that can be parsed by a PDR (and are therefore context free) that never has to look ahead more than a fixed number, say k, of symbols in the input are called $LR(k)$ *languages* (LR stands for $Look$ ahead $Right$). The $LR(k)$ languages are the deterministic context free languages. In fact, the deterministic languages are all $LR(1)$ *languages*. Languages of arbitrary length palindromic strings with no distinguishable central marker are clearly not $LR(k)$ languages, for any fixed k.

The preceding chapter closed with the observation that the language $\{a^i b^i : i \geq 1\}$ is not regular. It was promised that this would be proved in Chapter 6. Now we make a similar observation with respect to the context free languages. We argue that there are languages that are not context free, and cannot therefore be generated by a context free grammar, or recognised by an NPDR, and therefore must be either *type* 1 (*context sensitive*) or *type* 0 (*unrestricted*) languages (or perhaps neither).

In Chapter 6, we will see that both the "multiplication language" and the language:

$$\{a^i b^i c^i : i \geq 1\}, \quad \text{i.e. the set of all strings of the form } a\text{s followed by } b\text{s}$$
$$\text{followed by } c\text{s, where the number of } a\text{s} = \text{the number}$$
$$of \ b\text{s} = \text{the number of } c\text{s},$$

are not context free. Given that the language $\{a^i b^i : i \geq 1\}$ *is* context free, it may at first seem strange that $\{a^i b^i c^i : i \geq 1\}$ is not. However, this is indeed the case. Part of the next chapter is concerned with demonstrating this.

EXERCISES

For exercises marked "†", solutions, partial solutions, or hints to get you started appear in "Solutions to Selected Exercises" at the end of the book.

5.1. Given the following grammar:

$$S \to ABCA$$
$$A \to aA \,|\, XY$$
$$B \to bbB \,|\, C$$
$$X \to cX \,|\, \varepsilon$$
$$Y \to dY \,|\, \varepsilon$$
$$C \to cCd \,|\, \varepsilon$$

produce an equivalent Chomsky normal form grammar.
Note:remember that you first need to apply the rules for removing the empty string.

5.2.[†] $\{a^i b^j : i \geq 1,\ j = i\ or\ j = 0\}$ is a *deterministic CFL*. However, as mentioned earlier in this chapter, it can only be accepted by a deterministic PDR that does not always clear its *stack*. Provide a DPDR to accept the language and justify the preceding statement.

5.3. Argue that we can always ensure that an NPDR clears its stack and that NPDRs need have only one *halt state*.

5.4. Amend $M_3{}^d$, from earlier in the chapter (Figure 5.8), to produce *DPDRs* for the languages $\{a^i b^i c^j : i, j \geq 1\}$ and $\{a^i b^j c^j : i, j \geq 1\}$.
Note: in Chapter 6 it is assumed that the above two languages are deterministic.

5.5.[†] Justify by intuitive argument that $\{a^i b^j c^k : i, j \geq 1, i = j\ or\ j = k\}$ (from exercise 1 in Chapter 3) is a *non-deterministic CFL*.

5.6.[†] Design a program to simulate the behaviour of a DPDR. Investigate what amendments would have to be made to your program so that it could model the behaviour of an NPDR.

5.7. Design DPDRs to accept the languages $\{a^i b^j c^{i+j} : i,\ j \geq 1\}$ and $\{a^i b^j c^{i-j} : i,\ j \geq 1\}$.
Note: consider the implications of these machines in the light of the remarks about the computational power of the PDR, above.

6

Important Features of Regular and Context Free Languages

6.1 Overview

In this chapter we investigate the notion of *closure*, which in terms of languages essentially means seeing if operations (such as *union*, or *intersection*) applied to languages of a given class (e.g. *regular*) always result in a language of the same class. We look at some closure properties of the *regular* and *context free* languages. We next discover that the *Chomsky hierarchy* is a "proper" hierarchy, by introducing two theorems:

- the *repeat state theorem* for *finite state recognisers*

- the *uvwxy theorem* for *context free languages*.

We then use these theorems to show that:

- there are *context free* languages that are not *regular*

- there are *context sensitive* and *unrestricted* languages that are not *context free*.

A.P. Parkes, *A Concise Introduction to Languages and Machines*,
DOI: 10.1007/978-1-84800-121-3_6, © Springer-Verlag London Limited 2008

6.2 Closure Properties of Languages

In terms of the *Chomsky hierarchy*, the notion of *closure* applies as follows:

1. Choose a given type from the Chomsky hierarchy.

2. Choose some operation that can be applied to any language (or pairs of languages) of that type and which yields a language as a result.

3. If it can be demonstrated that when applied to each and every language, or pair of languages, of the chosen type, the operation yields a resulting language which is also of the chosen type, then the languages of the chosen type are said to be *closed under the operation*.

For example, take the set *union* operator, $\cup$, and the *regular* languages. If for *every* pair L_1, L_2, of regular languages the set $L_1 \cup L_2$ is also a regular language, then the regular languages would be *closed under union* (in fact they are, as we see shortly).

In the following two sections, we investigate various closure properties of the regular and context free languages, with respect to the operations *complement, union, intersection* and *concatenation*. As usual, we will be content with intuitive argument.

6.3 Closure Properties of the Regular Languages

6.3.1   Complement

As we discovered early on in the book, given some finite non-empty alphabet A, A^* denotes the *infinite* set of all possible strings that can be formed using symbols of that alphabet. Given this basis, a *formal language* was defined to be any subset of such a set. Take for example the language

$$R = \{a^i b^j c^k : i, j, k \geq 1\},$$

which is a subset of $\{a, b, c\}^*$.

Now, $\{a, b, c\}^* - R$ denotes the set of all strings that are in $\{a, b, c\}^*$ *but not in* R ("–" is called the *set difference* operator) i.e.

- all possible strings of *a*s and/or *b*s and/or *c*s except for those of the form "one or more *a*s followed by one or more *b*s followed by one or more *c*s".

$\{a, b, c\}^* - R$ is called the *complement* of R, and is, of course, a formal language, as it is a subset of $\{a, b, c\}^*$.

In general, then, if L is a language in A^*, $A^* - L$ is the *complement* of L.
We are now going to see that

- if L is a regular language, then so is $A^* - L$.

In other words, we will see that the regular languages are *closed under complement*. This is best demonstrated in terms of *deterministic finite state recognisers* (*DFSRs*). Our argument will be illustrated here by using the DFSR for R, the language described above, as an example.

We start with a DFSR that *accepts* R, as shown in Figure 6.1.

To the machine in Figure 6.1 we add a *null state* to make the machine completely deterministic. The resulting machine can be seen in Figure 6.2.

We take the machine in Figure 6.2 and make all its non-halt states into *halt states* and all its halt states into non-halt states, achieving the machine shown in Figure 6.3.

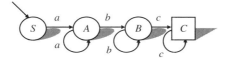

Figure 6.1 A DFSR for the language $R = \{a^i b^i c^k : i, j, k \geq 1\}$.

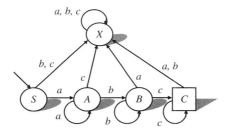

Figure 6.2 The DFSR from Figure 6.1 with a "null" state added to make it fully deterministic.

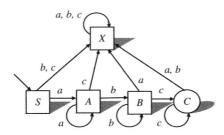

Figure 6.3 The "complement" machine of the DFSR in Figures 6.1 and 6.2.

The FSR in Figure 6.3 will reach a halt state only for strings in $\{a, b, c\}^*$ for which the original machine (Figure 6.1) would not. It thus accepts the *complement* of the language accepted by the original machine. Note that as the machine's start state is also a *halt* state, the final machine also accepts as a valid input the empty string (ε), as it should, since the original machine did not.

It should be clear that the same method could be applied to *any* DFSR. As we saw in Chapter 4, there is a general *equivalence* between DFSRs and regular languages. What applies to all DFSRs, then, applies to all regular languages.

6.3.2 Union

We now see that

- if L_1 and L_2 are regular languages, then so is $L_1 \cup L_2$.

  Consider again the DFSR for R, in Figure 6.1, and the DFSR for the language

$$\{a^i b^{2j} : i \geq 0, j \geq 1\},$$

which is shown in Figure 6.4.

Figure 6.5 shows a machine that is an amalgam of the machines from Figures 6.1 and 6.4. As can be seen in Figure 6.5, we have introduced an additional

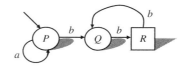

Figure 6.4 A DFSR for the language $\{a^i b^{2j} : i \geq 0, j \geq 1\}$.

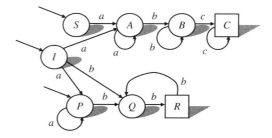

Figure 6.5 The non-deterministic FSR that is the union machine for the FSRs in Figures 6.1 and 6.4. Note: the machine has three start states (S, I and P).

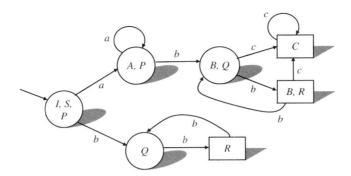

Figure 6.6 A deterministic version of the machine in Figure 6.5.

start state, labelled I, and for each outgoing arc from a start state of either original machine we draw an arc from state I to the destination state of that arc.

The machine in Figure 6.5 has three start states, but if we use our *subset algorithm* from Chapter 4 (Table 4.6) to make it deterministic, we obtain the FSR in Figure 6.6.

The machine in Figure 6.6 accepts the language:

$$\{a^i b^j c^k : i, j, k \geq 1\} \cup \{a^i b^{2j} : i \geq 0, j \geq 1\}, \quad \text{i.e. all strings accepted by either,}$$
$$\text{or both, of the original machines.}$$

You should again satisfy yourself that the suggested method could be applied to any pair of DFSRs.

6.3.3 Intersection

Now we establish that

- if L_1 and L_2 are regular languages, then so is $L_1 \cap L_2$

(the *intersection* of two sets is the set of elements that occur in both of the sets).

A basic law of set theory is one of *de Morgan's* laws, which is stated in Figure 6.7. Figure 6.8 shows an illustrative example to help to convince you that the rule is valid.

The law represented in Figure 6.7 tell us that the intersection of two sets is the same as the complement of the union of the complements of the two sets.

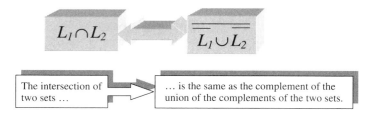

Figure 6.7 de Morgan's law for the intersection of two languages (sets).

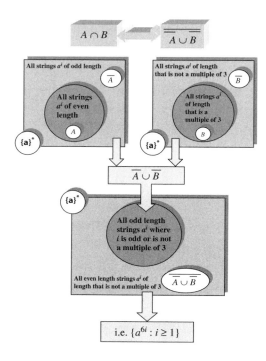

Figure 6.8 An illustration of de Morgan's law applied the intersection of the two languages $A = \{a^{2i} : i \geq 0\}$ and $B = \{a^{3i} : i \geq 1\}$.

Now, we have to do little to establish that if L_1 and L_2 are regular languages, then so is $L_1 \cap L_2$, since:

- we saw above how to construct the *complement* and

- we also saw how to create the *union*.

We can therefore take the two DFSRs for L_1 and L_2, say M_1 and M_2, and do the following:

a) we create the "complement machines" for each of M_1 and M_2

b) we create the "union machine" from the results of (a)

c) we create the "complement machine" from the result of (b).

We know that the complement of a regular language is always regular and the union of two regular languages is also always regular. The language recognised by the FSR; that results from carrying out steps (a) to (c) will therefore also be regular.

6.3.4 Concatenation

As two strings can be *concatenated* together, so can two languages. To concatenate two languages A and B together, we simply make a set of all of the strings which result from concatenating each string in A in turn, with each string in B. So, for example, if

$$A = \{a, ab, \varepsilon\}$$

and

$$B = \{bc, c\},$$

then A *concatenated to* B, which we write AB, or in this case $\{a, ab, \varepsilon\}\{bc, c\}$ would be

$$\{abc, ac, abbc, bc, c\}.$$

Analogously to the way we use the *superscript* notation for strings, so we can use the superscript notation for *languages*, so that A^3 represents the language A concatenated to itself, and the result of that concatenated to A again (i.e. AAA), or in this case (given the language A as defined above):

$$A^3 = \{a, ab, \varepsilon\}\{a, ab, \varepsilon\}\{a, ab, \varepsilon\}$$
$$= \{aa, aab, a, aba, abab, ab, \varepsilon\}\{a, ab, \varepsilon\}, \text{etc.}$$

Concatenation can be applied to infinite languages, too. For example, we could write

$$\{a^i bj : i, j \geq 1\}\{c^i : i \geq 1\},$$

to mean the set containing each and every string in the first set concatenated to each and every string in the second. In this case, the language described is R from earlier in this chapter, i.e.

$$\{a^i b^j c^k : i, j, k \geq 1\}.$$

The next thing we are going to show is that

- if L_1 and L_2 are regular languages, then so is $L_1 L_2$.

This is easier to show by using grammars than FSRs. Table 6.1 describes the combining of two regular grammars G_{13} and G_{14}, to produce a grammar, G_{15}, such that $L(G_{15}) = L(G_{13}) L(G_{14})$, i.e.

$$= \{a^i b^j x : i \geq 0, j \geq 1, x = b \text{ or } x = c\}\{a^i d^j : i, j \geq 1\}$$

$$= \{a^h b^i x a^j d^k : h \geq 0, i, j, k \geq 1, x = b \text{ or } x = c\}.$$

It should be clear that the method in Table 6.1 would work for any pair of regular grammars. Note that the method could also be used repeatedly, beginning with a grammar G, to create a grammar that generated $L(G)^n$, for any finite n.

Table 6.1 Combining two regular grammars to create a regular grammar that generates the concatenated language.

Two example grammars	
G_{13}	G_{14}
$\quad S \rightarrow aS \mid bB$	$\quad S \rightarrow aS \mid aA$
$\quad B \rightarrow b \mid c \mid bB$	$\quad A \rightarrow dA \mid d$
so that:	
$L(G_{13}) = \{a^i b^j x : i \geq 0, j \geq 1, x = b \text{ or } x = c\}$	$L(G_{14}) = \{a^i d^j : i, j \geq 1\}$

To produce a composite grammar that generates $L(G_{13})L(G_{14})$:

Step	Description	Result
1	Rename the non-terminals in the second grammar so that the two grammars have no non-terminals in common.	G_{14} is now: $\quad X \rightarrow aX \mid aA$ $\quad A \rightarrow dA \mid d$ with X as the start symbol
2	In the first grammar, replace each production of the form $x \rightarrow y$, where y is a single terminal, by a production $x \rightarrow yZ$, where Z is the start symbol of the second grammar.	G_{13} is now: $\quad S \rightarrow aS \mid bB$ $\quad B \rightarrow bX \mid cX \mid bB$
3	Put all the productions together. They now constitute one grammar, the start symbol of which is the start symbol of the first grammar. The resulting grammar generates the language that is the language generated by the first grammar concatenated to the language generated by the second.	Composite grammar, G_{15}: $\quad S \rightarrow aS \mid bB$ $\quad B \rightarrow bX \mid cX \mid bB$ $\quad X \rightarrow aX \mid aA$ $\quad A \rightarrow dA \mid d$

6.4 Closure Properties of the Context Free Languages

In this section, we investigate the closure properties of the *context free languages* (*CFLs*) in general. We find that the situation is not so straight forward for the CFLs as for the regular languages. We also find that there are differences between the closure properties of the CFLs in general, and the *deterministic CFLs*, the latter being described in Chapter 5 as those CFLs recognised by *deterministic PDRs*. For the CFLs to be closed under an operation means that the language resulting from the application of the operation to *any* CFL (or pair of CFLs, if the operation applies to two CFLs) is also a CFL. Moreover, for the *deterministic* CFLs to be closed under an operation means that the result must always be a *deterministic* CFL.

To support the arguments, we consider the following languages:

$$X = \{a^i b^i c^j : i, j \geq 1\}$$

and

$$Y = \{a^i b^j c^j : i, j \geq 1\}.$$

First of all, let us observe that the two languages are context free. In fact, X and Y are both *deterministic* CFLs, as we could design DPDRs to accept each of them, as you were asked to do in the exercises at the end of Chapter 5.

In any case, they are certainly context free, as we can show by providing CFGs for each of them. The two grammars, G_x and G_y, are shown in Table 6.2.

To simplify our constructions, the two grammars in Table 6.2 have no *non-terminals* in common.

6.4.1   Union

The first result is that the CFLs are closed under *union*, i.e.,

• if L_a and L_b are context free languages, then so is $L_a \cup L_b$.

Table 6.2 Two example grammars, G_x and G_y, to generate the languages $X = \{a^i b^i c^j : i, j \geq 1\}$ and $Y = \{a^i b^j c^j : i, j \geq 1\}$ respectively.

$G_x\ (L(G_x) = X)$	$G_y\ (L(G_y) = Y)$
$A \rightarrow BC$ {start symbol is A}	$D \rightarrow EF$ {start symbol is D}
$B \rightarrow aBb \mid ab$	$E \rightarrow aE \mid a$
$C \rightarrow cC \mid c$	$F \rightarrow bFc \mid bc$

Table 6.3 Constructing a CFG to generate the union of two CFLs. The example column features grammars G_x and G_y from Table 6.2.

Step	Description	Example
1	Arrange it so the two grammars have no non-terminals in common.	G_x and G_y are already like this
2	Collect all the productions of both grammars together.	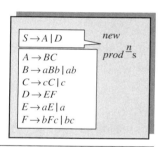
3	Add the two productions $$S \to P \mid Q,$$ where S is a new non-terminal and P and Q are the start symbols of the two grammars. S is now the start symbol of the composite grammar.	

Suppose that for L_a and L_b there are CFGs, G_a and G_b, respectively. Then to produce a grammar that generates the union of L_a and L_b, we carry out the process described in Table 6.3 (the example on the right of Table 6.3 uses the two grammars G_x and G_y from Table 6.2).

Referring to Table 6.3, you should be able to see that the grammar resulting from step 3 generates all of the strings that could be generated by one grammar or the other. Step 1 is necessary so that any derivation uses only productions from one of the original grammars. If G_a and G_b are the original grammars, the resulting grammar generates $L(G_a) \cup L(G_b)$. You can also see that if the two starting grammars are context free, the resulting grammar will also be context free.

The *context free languages* are therefore *closed under union*.

Now we focus on our particular example. The new grammar that we obtained from G_x and G_y generates $X \cup Y$, where X and Y are the languages above. This set, which will be called Z, is

$$Z = X \cup Y = \{a^i b^j c^k : i, j, k \geq 1, i = j \text{ or } j = k\}.$$

Now, while this is a CFL, and represents the union of two *deterministic* CFLs, it is not itself a deterministic CFL. To appreciate this, consider a PDR to accept it.

Such a machine could not be deterministic, as when it was reading the as in the input it would have to assume that it was processing a string in which the number of as was equal to the number of bs, which may not be the case (it may be processing a string in which the number of bs was equal to the number of cs). There would therefore be a need for *non-determinism*. For a class of languages to be closed under an operation, the result of every application of the operation must also be a language of that class. So our example (and there are many more) shows us that

- the *deterministic context free languages* are not *closed under union*.

However, as we showed above, in general, the CFLs *are* closed under union.

6.4.2 Concatenation

In this section, we see that

- if L_a and L_b are context free languages, then so is $L_a L_b$.

Remember that, if X and Y are languages, then XY denotes the language resulting from the *concatenation* of each and every string in X with each and every string in Y. Table 6.4 shows a simple method for taking two CFGs, G_a and G_b, and producing from them a new CFG, G, which generates $L(G_a)L(G_b)$. As before, we use our grammars G_x and G_y , from Table 6.2, to illustrate.

Table 6.4 Constructing a CFG to generate the concatenation of two CFLs. The example features grammars G_x and G_y from Table 6.2.

Step	Description	Example
1	Arrange it so the two grammars have no non-terminals in common.	G_x and G_y are already like this
2	For each production in the first grammar with S (the start symbol of the grammar) on its left-hand side, put R (the start symbol of the second grammar) at the end of the right-hand side of that production.	New production created: $A \to BC\underline{D}$ (the only one created, since G_x has only one production with A, its start symbol, on the left-hand side)
3	Add the rest of the first grammar's productions, and all of the second grammar's productions to the productions created in step 2. The start symbol of the first grammar is the start symbol of the new grammar.	New grammar:

The new grammar created in Table 6.4 ensures that any *terminal string* that could be derived by the first grammar is immediately followed by any terminal string that the second grammar could produce. Therefore, the new grammar generates the language generated by the first grammar concatenated to the language generated by the second grammar. The resulting grammar in Table 6.4 generates the set XY, such that

$$XY = \{a^i b^i c^j a^s b^t c^t : i, j, s, t \geq 1\}.$$

The method of construction specified in Table 6.4 could be applied to any pair of context free grammars, and so *the context free languages are closed under concatenation.*

However, the *deterministic* CFLs are not closed under concatenation. This is not demonstrated by our example, because XY happens to be deterministic in this case. The two *deterministic* CFLs

$$\{a^i b^j : i \geq 1, j = 0 \ or \ j = i\}$$

and

$$\{b^i c^j : i = 0 \ or \ i = j, j \geq 1\}$$

concatenated together result in a language which is a *non-deterministic* CFL. You are asked in the exercises to argue that this is indeed the case.

Our concatenation result for the deterministic CFLs is thus:

- the *deterministic context free languages* are not *closed under concatenation.*

6.4.3   Intersection

Thus far, we have seen that the closure situation is the same for the CFLs as for the regular languages, with respect to both union and concatenation. However, in terms of *intersection*, the CFLs differ from the regular languages, in that the CFLs are not closed under intersection. To appreciate this, consider the two sets we have been using in this section, i.e.

$$X = \{a^i b^i c^j : i, j \geq 1\}$$

and

$$Y = \{a^i b^j c^j : i, j \geq 1\}.$$

What is the intersection $(X \cap Y)$ of our two sets? A little thought reveals it to be the set of strings of as followed by bs followed by cs, in which the number of as equals the number of bs *and* the number of bs equals the number of cs, i.e.,

$$X \cap Y = \{a^i b^i c^i : i \geq 1\}.$$

This language was stated to be not context free at the end of the last chapter and this will be demonstrated later in this chapter. Moreover, both X and Y are *deterministic* CFLs, so this counter-example shows us that:

- both the *deterministic* and the *non-deterministic context free languages* are not *closed under intersection*.

6.4.4 Complement

We can argue that it is also the case that the CFLs are not closed under *complement* (i.e. that the complement of a CFL is not necessarily itself a CFL). You may recall that above we used de Morgan's law (Figure 6.7) to show that the regular languages were closed under intersection.

We can use the same law to show that the CFLs cannot be closed under complement. If the CFLs *were* closed under complement, the above law would tell us that they were closed under intersection (as we showed that they were closed under *union* earlier on). However, we have already shown that the CFLs are not closed under intersection, so to show that this is the case is absurd. We must therefore reject any assumption that the CFLs are closed under complement.

The above argument is our first example of proof by a technique called *reductio ad absurdum*. We simply assume the logical negation of the statement we are trying to prove, and on the basis of this assumption follow a number of logically sound arguments that eventually lead to a nonsensical conclusion. Since the steps in our argument are logical, it must be our initial assumption that is false. Our initial assumption is false, but it is actually the negation of the statement we are trying to prove. Therefore, the statement we are trying to prove is indeed true.

In this case, what our *reductio ad absurdum* proof shows us is that

- the *context free languages* are not *closed under complement*.

This result is very important, at least with respect to the *pushdown recogniser*. Imagine you are writing a *parser* for a programming language that you have designed. Suppose you implement your parser to be equivalent in power to a non-deterministic pushdown recogniser. It turns out, unknown to you, that your language is one for which the complement is not a context free language. Now, this is not likely to cause problems for your parser if its input consists entirely of *syntactically correct* programs. This is highly unlikely, I'm sure you will agree, if you have ever done any programming! However, there will be one (or more) syntactically incorrect "programs" for which your parser will not be able to make any decision at all (it will probably enter an infinite loop). If, for every string we input to the parser, the parser (which remember is no more powerful than a PDR) could output an indication of whether *or not* the program is syntactically legal, the same parser could be used to accept the complement of our original language (i.e. we simply interpret the original parser's "yes" as a "no", and *vice versa*). However, the fact that the CFLs are not closed under complement tells us that for some languages a parser equivalent in power to a PDR will be unable to make the "no" statement. I suggest you re-examine the *acceptance conditions* for *non-deterministic PDRs* in the previous chapter (Table 5.5). Now you can probably see why there is no condition that dictates that the machine must *always* halt, only that it must always halt in the case of *sentences* of the language.

The issue of whether or not *abstract machines* eventually stop at some time during a computational process is central to computer science. However, we defer consideration of such issues until Part 2 of this book.

All is not lost, however, for two reasons. Firstly, if your programming language is a *deterministic* CFL, its complement will also be a deterministic CFL. Remember, at the end of Chapter 5 we saw that a deterministic CFL was a language for which a pushdown machine did not have to look more than one symbol ahead in the input. We saw that deterministic CFLs are called *LR* (1) *languages* for that reason. Well, in such cases there are only a finite number of symbols the machine expects to encounter when it looks ahead to the next one. They form two sets at each stage: the symbols the machine expects to see next, and the symbols it does not expect to see next. If at any stage, the machine looks ahead and finds a symbol which at that stage is *not* expected, the machine simply rejects the input as invalid (i.e., moves to a halt state that denotes *rejection*, as opposed to a halt state that denotes *acceptance*). We can make sure, then, that our machine always halts with "yes" or "no". We can then simply exchange the acceptance/rejection status of our machine's halt states, and we have a machine that accepts the complement of the original language. So,

- the *deterministic context free languages* are *closed under complement*.

6.5 Chomsky's Hierarchy is Indeed a Proper Hierarchy

In this section, we sketch out two fundamental theorems of formal languages. One is for the *regular languages*, though it actually applies to the FSR, and is called the *repeat state theorem*. The other, for the *context free* languages, is in some ways similar to this, and is called the *uvwxy theorem*. The usefulness of these theorems is that they apply to every single language in their class, and can therefore be used to show that a given language is not in the class. To appreciate this point, think about it like this: if every element in a set must possess a particular property, and we come across an object that does not have that property, then the object cannot belong to our set. If it is true that every bird has feathers, and we come across a creature that is featherless, then that creature is certainly not a bird.

If we use the *repeat state theorem* to show that a certain language cannot be regular, but we can show that the same language is context free, then we know that the regular languages are a *proper subset* of the context free languages. We already know that every regular language is context free, but then we will have also shown that there are context free languages that are not regular. Moreover, if we use the *uvwxy theorem* (considered later in this chapter) to show that a given language cannot be context free, and we then show that it is *type* 0, we then also know that the context free languages are a proper subset of the type 0 languages.[1] We have then established that *Chomsky's hierarchy* is a real hierarchy.

Now, some students experience problems in thinking in terms of proper subsets. This is because when you say to them "all xs are ys", they automatically assume that you also mean "all ys are xs". To demonstrate the silliness of this inference, consider *parrots* and *birds*. If I said to you "all parrots are birds", you would not then think "all birds are parrots". The set of all regular languages corresponds to the set of all parrots, and the set of all context free languages corresponds to the set of all birds. One set is *properly* contained in the other.

6.5.1 The "Repeat State Theorem"

Any FSR that has a *loop* on some path linking its start and halt states recognises an infinite language.

Consider the example FSR, M, in Figure 6.9.

M, in Figure 6.9, has 7 states. Consider any string acceptable to M that is greater than 6 symbols in length. That string must have resulted in the machine passing through one of its states more than once. After all, M has only 7 states, and it would need 8 states to accept a string of length 7, if no state was to be visited more than once. Consider, as a suitable example, the string *abbacaab*. This string is 8 symbols in length. The sequence of states visited during the acceptance of this string is indicated in Figure 6.10.

---

[1] We omit the *context sensitive* (type 1) languages from this discussion. Only the ε-free CFGs are context sensitive, as context sensitive grammars are not allowed to have any ε productions.

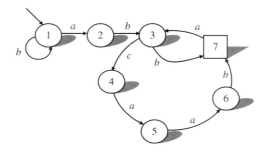

Figure 6.9 An FSR, M, that contains a *loop* of states.

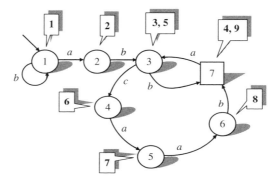

Figure 6.10 The order in which the states of M are visited when the input string
is *abbacaab*.

Let us choose any state that M visited more than once in accepting our string.
Choose state 3. State 3, as the diagram shows, was the third and fifth state visited.
Consider the substring of *abbacaab* that caused M to visit state 3 twice, i.e. *ba*.
Where *ba* occurs in *ab<u>ba</u>caab*, we could copy it *any* number of times and we would
still obtain a string that M would accept. As an example, we will copy it 4 times:
ab<u>babababa</u>caab. Furthermore, we could omit *ba*, giving *abcaab*, a string that is
also accepted by our machine.

The other state that was visited twice in the acceptance of *abbacaab* was state 7
(the halt state). We now choose this as our example repeated state. The substring
of *abbacaab* that caused state 7 to be visited twice was *acaab*. As before, our
machine M would accept the original string with no copies of *acaab* in it: *abb*. It
would also accept the original string with any number of copies inserted after the
first occurrence, for example 3 copies: *ab<u>bacaabacaabacaab</u>*

To appreciate the generality of the above, consider any FSR that accepts a string of length greater than or equal to the number of states it has. As we said before, in accepting such a string one or more of its states must have been visited twice. We consider any of those "visited-twice" states. The string that was accepted can be expressed as three concatenated strings:

- a possibly empty *first part*, representing the states visited *before* the machine got to its "visited twice" state,

- a non-empty *second part*, representing the part of the string that caused the machine to visit the "visited-twice" state again. This part of the string can be copied 0 or more times in between the *first* and *third* parts, and the resulting string will also be acceptable to the machine,

- a possibly empty *third part*, representing the states visited *after* the machine has left the "visited-twice" state loop

In our examples for the string *abbacaab*, above, when *state* 3 was the chosen "*visited-twice*" *state*:

- *first part was ab,*

- *second part was ba,*

- *third part was caab.*

   Alternatively, when *state* 7 was the chosen "*visited-twice*" *state*:

- *first part was abb,*

- *second part was acaab,*

- *third part was empty.*

In many cases, *first part* may be empty, if the loop began at the start state. In other cases, both *first* and *third parts* may be empty, if the loop encompasses every state in the machine.

   The repeat state theorem is expressed formally in Table 6.5.

   As was established in Chapter 4, any language accepted by a DFSR is a regular language. Therefore, our theorem applies to any *infinite* regular language.

Table 6.5 The repeat state ("*vwx*") theorem for regular languages.

For any deterministic FSR, M:

If M accepts a string z, such that $|z| \geq n$, where n is the number of states in M, then z can be represented as three concatenated substrings
   $z = vwx$
in such a way that all strings vw^ix, for all $i \geq 0$, are also accepted by M.

6.5.2   A Language that is Context Free but not Regular

Now, as promised at the end of Chapter 4, we turn our attention to the language

- $\{a^i b^i : i \geq 1\}$

We know that this language is context free, because it is generated by the context free grammar G_3, of Chapter 2:

$$S \rightarrow aSb \,|\, ab.$$

The question is, could it also be a regular language? We shall now use our new theorem to show that it cannot be a regular language, using the technique of *reductio ad absurdum* introduced earlier in this chapter.

Our assumption is that $\{a^i b^i : i \geq 1\}$ *is* a regular language. If it is regular it is accepted by some FSR, M. M can be assumed to have k states. Now, k must be a definite number, and clearly our language is infinite, so we can select a sentence from our language of the form $a^j b^j$, where $j \geq k$. The repeat state theorem then applies to this string. The repeat state theorem applied to our language tells us that our sentence $a^j b^j$ can be split into three parts, so that the middle part can be repeated any number of times within the other two parts, and that this will still yield a string of as followed by an equal number of bs. But how can this be true? Consider the following:

- the "*repeatable substring* " cannot consist of as followed by bs, as when we repeated it we would get as followed by bs followed by as... Our machine would accept strings that are not in the language $\{a^i b^i : i \geq 1\}$.

This means that the repeatable substring must consist of as alone or bs alone, However:

- the repeatable substring cannot consist of as alone, or when we repeated it more than once, the as in our string would outnumber the bs, and our machine would accept strings not in the language $\{a^i b^i : i \geq 1\}$.

A similar argument obviously applies to a repeatable substring consisting entirely of bs.

Something is clearly amiss. Every statement we have made follows logically from our assumption that $\{a^i b^i : i \geq 1\}$ is a regular language. We are therefore forced to go right back and reject the assumption itself. $\{a^i b^i : i \geq 1\}$ is not a regular language.

Our result tells us that an FSR could not accept certain strings in a language like $\{a^i \ b^i : i \geq 1\}$ without accepting many more that were not supposed to be acceptable. We simply could not constrain the FSR to accept only those strings in the language. That is why an FSR can accept a language such as $\{a^i\,b^j : i,j \geq 1\}$, as you will find out if you design an FSR to do so. However, the language $\{a^i\,b^i : i \geq 1\}$, which is a *proper subset* of $\{a^i\,b^j : i,j \geq 1\}$ is beyond the computational abilities of the FSR.

One interesting outcome of discovering that $\{a^i\,b^i : i \geq 1\}$ is not regular is that we can use that fact to argue that a language such as Pascal is not regular. In a sense, the set of Pascal <compound statement>s is of the form:

$$\{\text{begin}^i\,x\,\text{end}^i : x \text{ is a sequence of Pascal } <statement>s\}.$$

The repeat state theorem tells us that many strings in this part of the Pascal language cannot be accepted by an FSR unless the FSR accepts many other strings in which the number of *begin*s does not equal the number of *end*s. By extension of this argument, the *syntax* of Pascal cannot be represented as a regular grammar. However, as mentioned in Chapter 3, the syntax of Pascal can be expressed as a *context free* grammar. The same applies to most programming languages, including, for example, Lisp, with its incorporation of strings of arbitrary length "balanced" parentheses.

You are asked in the exercises to provide similar justifications that certain other context free languages cannot be regular. It is thus clear that the regular languages are properly contained in the class of context free languages. In fact, we have shown *two* things about Chomsky's hierarchy:

1. the *regular languages* are a *proper subset* of the CFLs in general, and, more specifically,

2. the *regular languages* are a *proper subset* of the *deterministic* CFLs.

Point 2 is demonstrated by considering that $\{a^i\,b^i : i \geq 1\}$ is not regular, but is a CFL and is also deterministic (as we saw in Chapter 5). So there is at least one language that is a deterministic CFL that is not regular. In fact, none of the deterministic CFLs we considered when establishing the *closure* results for CFLs in this chapter are regular. You might like to convince yourself of this, by using the repeat state theorem.

6.5.3   The "uvwxy" Theorem for Context Free Languages

The theorem described in this section is a context free language counterpart of the *repeat state theorem* described above. However, we establish our new theorem by using properties of context free grammars and *derivations* rather than the corresponding abstract machines (*PDRs*). You may find the argument difficult to follow. It is more important that you appreciate the implications of the theorem, rather than understand it in great detail.

Let us assume that we have an ε-free CFG that is in *Chomsky normal form* (*CNF*). That we can assume that any CFG is so represented as justified in Chapter 5.[2] As an example, we will use a CNF version of $G_3(S \rightarrow aSb \mid ab)$, which we will simply call G:

$$S \rightarrow AX \mid AB$$

$$X \rightarrow SB$$

$$A \rightarrow a$$

$$B \rightarrow b.$$

G has four *non-terminal* symbols. This means that if we construct a derivation tree for a sentence in which there is a path from S with more than four non-terminals on it, one of the non-terminals will appear more than once.

Now, consider the *derivation tree* for the sentence a^4b^4, shown in Figure 6.11 and labelled $DT1$.

In particular, we focus on the path in $DT1$ proceeding from the top of the tree to the non-terminal marked "*low X* ", in $DT1$. Sure enough, this path features a repeated non-terminal. In fact, both S – three times – and X – also three times – are repeated on this path. We can choose any pair of the repeated non-terminals, so we will choose the two instances of the non-terminal X that have been labelled "*low X* " and "*high X* ". Five *substrings*, that *concatenated* together make up the sentence, have been identified and labelled in Figure 6.11 as follows:

u this is the string derived in the part of the tree from S to the *left* of *high X*, and is the string *aa*

v this is the string derived from *high X* to the *left* of *low X*, and is the string *a*

w this is the string derived from *low X*, i.e. *abb*

x this is the string derived from *high X* to the *right* of *low X*, i.e. *b*

y this is the string derived from S to the *right* of *high X*, and is *b*.

Concatenating all the above strings in order, i.e. *uvwxy*, gives us our sentence a^4b^4.

Careful thought may now be required. The derivation from *high X* in $DT1$ is simply the application of a sequence of productions, the first having X as its left-hand side. G is a CFG, so the derivation we did from *high X* in $DT1$ could be repeated from *low X*. We would still have a sentence, as the whole derivation started with S, and a terminal string is derived from *high X* in $DT1$.

---

[2] We leave out the single ε production, if it exists, and concentrate only on the ε-free part of the grammar.

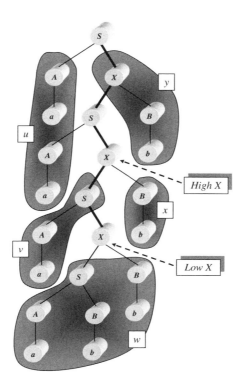

Figure 6.11 Derivation tree $DT1$ for the string a^4b^4, showing the characteristic context free $uvwxy$ form.

Doing the above gives us $DT2$, as shown in Figure 6.12. $DT2$ is the derivation tree for the sentence a^5b^5. As $DT2$ shows, the way we have obtained the new derivation ensures that we now have *two copies* of the substrings v and x, where there was one copy of each, respectively, in the $DT1$ sentence. Expressed in an alternative way, we have:

$$uv^2wx^2y(= a^5b^5).$$

What we have just done could be repeated, i.e. we could do the *high X* derivation we did in $DT1$ from the new *low X* near the bottom of $DT2$, in Figure 6.12. Then we would have uv^3wx^3y. Moreover, we could repeat this process indefinitely, i.e. we can obtain uv^iwx^iy, for all $i \geq 1$.

Finally, referring back to $DT1$ (Figure 6.11), the derivation originally carried out from *low X* could have been done from *high X* instead. If we do this, we obtain $DT3$, as shown in Figure 6.13. $DT3$ is the derivation tree for the sentence a^3b^3.

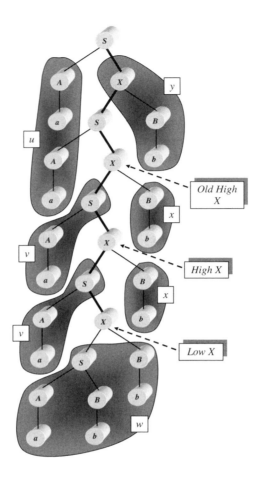

Figure 6.12 Derivation tree $DT2$ for the string a^5b^5, showing the uv^2wx^2y form
of $DT1$ in Figure 6.11.

As can be seen from Figure 6.13, this time there are only three of our substrings,
namely u, w and y. We have $uwy = a^3b^3$.

As stated in Chapter 2, for any string, s, $s^0 = \varepsilon$, the *empty string*, and for any
string z, $z\varepsilon = z$ and $\varepsilon z = z$. uwy can therefore be represented as uv^0wx^0y.

The constructions used to obtain the three trees, $DT1$–3 (Figures 6.11
to 6.13), and our accompanying discussion tell us that, for the language $L(G)$,
i.e. $\{a^ib^i : i \geq 1\}$, we have found a string, a^4b^4, which is such that:

a^4b^4 can be represented as concatenated substrings u, v, w, x and y, such that
$uvwxy = a^4b^4$, and for all $i \geq 0$, uv^iwx^iy is in $L(G)$.

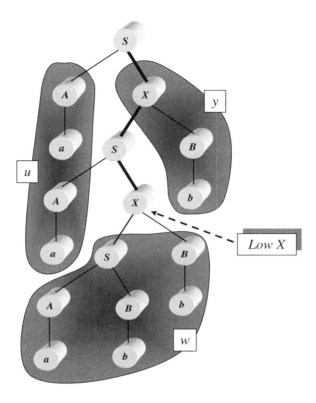

Figure 6.13 Derivation tree $DT3$ represents the *uwy* form of $DT1$ (Figure 6.11).

We initially chose the two Xs called *high X* and *low X* in $DT1$. However, any other equal pair of non-terminals on the path could have been chosen and a similar argument could have been made. For example, we could have chosen two of the repeated S nodes. A possible arrangement in this case is shown in $DT1a$, for which see Figure 6.14.

In Figure 6.14. there is no string u to the left of the top S down to *high S* (as *high S is* the top S), and similarly there is no string y to its *right*, both u and y are empty, i.e. $u = y = \varepsilon$. So we now have:

$$u = \varepsilon,$$
$$v = aa,$$
$$w = aabb,$$
$$x = bb, \text{and}$$
$$y = \varepsilon.$$

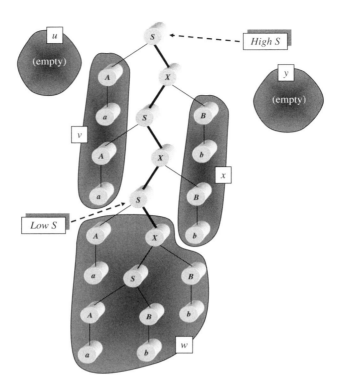

Figure 6.14 Derivation tree $DT1a$, an alternative $uvwxy$ form for $DT1$ (Figure 6.11).

This time, if we duplicated the original *high S* derivation from *low S* in Figure 6.14, we would have uv^2wx^2y, i.e., ε concatenated to *aa* concatenated to *aa* concatenated to *aabb* concatenated to *bb* concatenated to *bb* concatenated to ε, which is a^6b^6. The next time we repeated it we would obtain uv^3wx^3y, i.e. a^8b^8, and so on. If we duplicated the *low S* derivation from *high S*, we would obtain uv^0wx^0y. This, since all of the substrings except for *w* is empty, simply leaves us with *w*, i.e. *aabb*. So again, our statement above:

> a^4b^4 can be represented as concatenated substrings *u*, *v*, *w*, *x* and *y*, such that $uvwxy = a^4b^4$, and for all $i \geq 0$, uv^iwx^iy is in $L(G)$,

is true for this case also, as long as we permit *u* and *y* to be empty.

How general is the *uvwxy* type construction? It can be argued to apply to *any* context free grammar that generates a sentence long enough so that its derivation tree contains a path from S that contains two (or more) occurrences of one non-terminal symbol. Any CFG that generates an *infinite* language *must* be able to

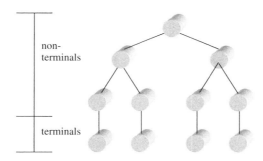

Figure 6.15 The shallowest derivation tree for a sentence that is four symbols long produced by Chomsky normal form (CNF) productions.

generate a derivation tree that contains a path like this. Such a CFG, as we showed in Chapter 5, can have its ε-free part converted into CNF. If there are n non-terminals in a CNF grammar, by the time we generate a sentence that is greater in length than 2^{n-1} (*numerically* speaking), we know we have a path of the type we need. To appreciate this, suppose that a CNF grammar has three non-terminals, i.e. $n = 3$. The shallowest tree for a sentence of length *equal* to 2^{n-1} (4) is shown in Figure 6.15.

Notice that the longest paths in the tree in Figure 6.15 have length, or number of *arcs*, 3, and, since the last *node* on the path would be a terminal, such paths have only 3 non-terminals on them. Such paths are not long enough to ensure that they feature repeated non-terminals. Our hypothetical grammar has 3 non-terminals, so to be absolutely sure that our tree contains a path of length greater than 3 we must generate a sentence of length *greater than* 4. You may need to think about this, perhaps drawing some possible extensions to the above tree will convince you. Remember, the grammar is assumed to be in CNF. You should also convince yourself that the 2^{n-1} argument applies to any CNF grammar.

It is not always necessary to generate a long sentence to obtain our "repeated non-terminal path". Our grammar G can be used to create derivation trees for sentences of length less than 8 (2^3, since G has 4 non-terminals) that have paths of the type we require. However, here we are interested in defining conditions that apply in general, which requires us to be *sure* that we can obtain a path of the type required.

Let G stand for *any* ε-free context free grammar. We now know that if we generate a sentence of sufficient length the derivation tree for that sentence will have a path, starting from the S at the root of the tree, upon which the same non-terminal, say X, appears twice or more. We choose two occurrences of X on that path, and call one *high* X and one *low* X. The sentence can be split into 5 concatenated substrings as before (use the derivation trees $DT\,1$–3 and $DT\,1a$ in Figures 6.11 to 6.14 for illustration, if you wish):

- u the string derived from S at the root and to the left of *high* X. This string may be empty, either because $S = high\,X$ (as in $DT1a$), or because our path proceeds directly down the left of the tree.

- v the string derived from *high* X to the *left* of *low* X. This string may be empty if the path from *high* X to *low* X is on the left. However, if this string is empty,

- x (see below) will not be. As we assume a CNF grammar, *high* X must expand initially to *two* non-terminals. Either (a) one of these is *low* X, or (b) one of these must eventually expand into two non-terminals of which one is *low* X. The other one expands into part of v or x, depending on the particular tree.

- w the string derived from *low* X. This string will not be empty, as the grammar is ε free.

- x as for v but derived from *high* X to the *right* of low X. The same argument as for v applies here. In particular, x cannot be empty if v is.

- y as for u but to the *right* of *high* X. Also may be empty, for analogous reasons to those given above, for u.

As was shown by our example derivation trees $DT\,1$–3, above, we can always repeat the "deriving what was derived from the *high* X in the original tree from *low* X" routine as many times as we wish, thus obtaining the characteristic uv^iwx^iy form, for all $i \geq 1$. Furthermore, we can derive what was derived from *low* X in the original tree from *high* X, giving us the uv^0wx^0y case.

Enough has been said to demonstrate the truth of the theorem that is formally stated in Table 6.6.

6.5.4 $\{a^ib^ic^i : i \geq 1\}$ is not Context Free

As for the *repeat state theorem* for regular languages, we can use our *uvwxy theorem* to show that a given infinite language *cannot* be context free. As promised at the end of Chapter 5, we now demonstrate that the language

$$\{a^ib^ic^i : i \geq 1\}$$

is not context free.

Table 6.6 The "*uvwxy*" theorem for context free languages.

For any infinite CFL, L :
Any sufficiently long sentence, z, of L can be represented as five concatenated substrings u, v, w, x, and y, such that $z = uvwxy$, and uv^iwx^iy is also in L, for all $i \geq 0$.

The proof is very simple. Again, we use *reductio ad absurdum* as our technique. We assume that $\{a^i \, b^i \, c^i : i \geq 1\}$ *is* a CFL. The language is clearly infinite, so the *uvwxy theorem* applies to it. Our sufficiently long sentence is going to be one greater in length than 2^{n-1}, where n is the number of non-terminals in a Chomsky normal form grammar that generates our language.

We can assume that our sufficiently long sentence is $a^k \, b^k \, c^k$, for some k, where $k > 2^{n-1}$. The theorem tells us that this sentence can be represented as five concatenated substrings *uvwxy*, such that $uv^i \, wx^i \, y$ are also in $\{a^i \, b^i \, c^i : i \geq 1\}$ for all $i \geq 1$. However, how are substrings of $a^k \, b^k \, c^k$ to be allocated to u, v, w, x and y in this way?

- Neither v nor x can be strings of a s and b s, or b s and c s, or a s, b s and c s, as this would result in symbols being in the wrong relative order when v and x were repeated.

So this means that the repeatable strings v and x must consist of strings which are either all as, all bs or all cs. However,

- this would also be inappropriate, as even if, say, v was aaa and x was bbb, when v and x were repeated, though the number of a s would still equal the number of bs, the number of each would exceed the number of c s. A similar argument can be used to show that no other allocation of substrings to v and x would work.

We must again reject our original assumption. $\{a^i \, b^i \, c^i : i \geq 1\}$ is not a context free language.

In fact, $\{a^i \, b^i \, c^i : i \geq 1\}$ is a *context sensitive* language, since it is generated by the following *context sensitive grammar*, G_{16}:

$$S \to aSBC \,|\, aBC$$
$$CB \to BC$$
$$aB \to ab$$
$$bB \to bb$$
$$bC \to bc$$
$$cC \to cc,$$

as you might like to justify for yourself.

The arguments in the following section may require some effort to appreciate. The section can be omitted, if required, though the result, as stated in the title of the section, should be appreciated.

6.5.5 The "Multiplication Language" is not Context Free

Now we consider one final refinement to our *uvwxy* theorem. We focus on a derivation tree for a sentence of length greater than 2^{n-1}, as described above.

We choose the longest path (or one of them, if there are several). Now, as our *high X* and *low X*, we pick the *lowest two repeated nodes on this path*, a constraint that was not enforced in our earlier examples. The string derived from *high X* (i.e. the one we call *vwx*) is derived by a tree with *high X* as its root in which the path that passes through *low X* is the longest path. We selected the two *X* s because they were the lowest repeated nodes on our longest path, therefore *high X* and *low X* are the *only* repeated nodes on our "sub-path". Therefore, the path must have maximum length $n + 1$, with the last node on the path being a terminal. Furthermore, the longest string that could be generated from *high X* (the one we call *vwx*) must be such that its length is less than or equal to 2^n (formally, $|vwx| \leq 2^n$).

We have established that for any sentence, z, such that $|z| > 2^{n-1}$, where there are n non-terminal symbols in a Chomsky normal form grammar that generated z, we can define the *uvwxy* substrings so that $|vwx| \leq 2^n$. To see why this can be useful, we consider the "multiplication language" introduced at the end of Chapter 2, i.e.

$$\{a^i b^j c^{i \times j} : i, j \geq 1\},$$

which was generated by our *type* 0 *grammar*, G_4(towards the end of Chapter 2). It was stated, also at the close of Chapter 2, that the above language is not context free. We now use the *uvwxy* theorem, along with our newly established result obtained immediately above, to demonstrate this.

Again, we use *reductio ad absurdum*. We assume that $\{a^i b^j c^{i \times j} : i, j \geq 1\}$ *is a* context free language. This means that the *uvwxy* theorem applies to a sufficiently long string. Consider a string, z from our language. $z = a^{2n} b^{2n} c^{(2n)^2}$, where n is the number of non-terminals in a Chomsky normal form grammar for the language. Obviously, the length of z is greater than 2^{n-1}. We can then, as shown in the preceding paragraph, establish a *uvwxy* form for this sentence, such that $|vwx| \leq 2^n$ (remember, n is the assumed number of non-terminals). Now, by a similar argument to that applied in the case of the language $\{a^i b^i c^i : i \geq 1\}$ above, we can certainly justify that v and x *cannot be mixtures of different terminals*. This means that v must consist either entirely of *a*s or entirely of *b*s, and x must be *c*s only (otherwise when we increase the number of *a*s or *b*s for v^i, the *c*s would not increase accordingly). Now,

- suppose v is a string of *a*s. Then, since x must consist of *c*s only, w must include all of the *b*s. This cannot be so, as both v and x are non-empty, so $|vwx|$ exceeds 2^n (since there are 2^n *b*s in w alone)

So v cannot be a string of *a*s then, which means that:

- v must be a string of *b*s. Now, suppose $v = b^k$, for some $k \geq 1$. Then x must be $c^{2 \times k}$. This is because each time $uv^i wx^i y$ inserts a new copy of v (b^k) into the

bs, it inserts a new copy of x into the cs. To ensure that the number of cs is still the same as the number of as times the number of bs, we must insert 2^n cs (2^n being the number of as) for each of the k bs. However, again this cannot be so, because if $|x| = 2^n \times k$, then $|vwx|$ is certainly $> 2^n$.

If the language had been context free, we could, for our sufficiently long sentence, have found a $uvwxy$ form such that $|vwx| \leq 2^n$. We assumed the existence of such a form and it led us to a contradiction. We are therefore forced, once more, to reject our original assumption and conclude that $\{a^i\, b^j\, c^{i\,\times j} : i, j \geq 1\}$ is not a context free language.

The "length constraint", i.e., that the vwx in the $uvwxy$ form is such that $|vwx| \leq 2^n$, where n is the number of non-terminals in a CNF grammar, makes the $uvwxy$ theorem especially powerful. It would have been difficult to establish that $\{a^i b^j c^{i \times j} : i,\ j \geq 1\}$ is not a context free language by arguing simply in terms of $uvwxy$ patterns alone. You may wish to try this to appreciate the point.

6.6 Preliminary Observations on the Scope of the Chomsky Hierarchy

The fact that $\{a^i b^j c^{i \times j} : i,\ j \geq 1\}$ is not context free suggests that we need more powerful machines than *pushdown recognisers* to perform *computations* such as multiplication. This is indeed the case. However, the abstract machine that *is* able to multiply arbitrary length numbers is not that far removed from the pushdown machine. The study of the machine, the *Turing machine*, will be a major feature of the second part of this book. It also turns out that the Turing machine is the recogniser for the context sensitive (type 1) and unrestricted (type 0) languages, which is how we introduce it in the next chapter.

In the preceding sections we have seen that there is indeed a hierarchy of *formal languages*, with each class being properly contained within the class immediately above it. The question arises: are the type 0 languages a *proper subset* of the set of all formal languages? As we said earlier, a formal language is simply any set of strings. Much later in this book (Chapter 11), we will see that there are indeed formal languages that are outside the class of type 0 languages, and thus cannot be generated by any grammar specified by the Chomsky hierarchy.

EXERCISES

For exercises marked "†", solutions, partial solutions, or hints to get you started appear in "Solutions to Selected Exercises" at the end of the book.

6.1. Produce a *DFSR* to accept the language $\{a^{4i} : i \geq 1\}$, then use the *complement* construction described earlier in the chapter to produce a machine that accepts the language $\{a^{4i+j} : i \geq 0, 1 \leq j \leq 3\}$.

6.2. As mentioned in the text above, both $L_1 = \{a^i b^j : i \geq 1, j = 0 \text{ or } j = i\}$ and $L_2 = \{b^i c^j : i = 0 \text{ or } i = j, j \geq 1\}$, are *deterministic CFLs*, but the language $L_1 L_2$ is not. Justify this.

6.3.† As expressed above, the first sentence of the section on the *repeat state theorem* is not generally true. Rewrite the sentence so that it is.

6.4.† Use the *repeat state theorem* to show that $\{a^i b^j c^{i+j} : i, j \geq 1\}$ and $\{a^i b^j c^{i-j} : i, j \geq 1\}$, both shown to be *deterministic* by exercise 7 of Chapter 5, are not *regular* languages.

6.5.† Use the *uvwxy theorem* to show that $\{xx : x \in \{a, b\}^*\}$ is *not context free*. (In $\{xx : x \in \{a, b\}^*\}$ each string consists of an arbitrary mixture of *a*s and/or *b*s of even length (or empty) and the second half of the string is an exact copy of the first, e.g. *aaabbbaaaabbba*.)

7
Phrase Structure Languages and Turing Machines

7.1 Overview

This chapter introduces the *Turing machine* (*TM*), an abstract machine for language recognition and computation.

We examine TMs as language recognisers. You will appreciate that:

- TMs can do anything that *Finite State Recognisers* (*FSRs*) and *Pushdown Recognisers* (*PDRs*) can do

- TMs are more powerful than *non-deterministic PDRs*

- TMs are the recognisers for the *phrase structure* (*type* 0) languages in general.

7.2 The Architecture of the Turing Machine

The *abstract machine* we consider in this chapter, and for much of Part 2 of this book, is called the *Turing machine* (*TM*). Alan Turing was the English mathematician who first described such a machine in 1936. A TM is essentially a finite state machine, like a *finite state recogniser*, but with some very simple additional facilities for input and output, and a slightly different way of representing the input sequences.

A.P. Parkes, *A Concise Introduction to Languages and Machines*,
DOI: 10.1007/978-1-84800-121-3_7, © Springer-Verlag London Limited 2008

7.2.1 "Tapes" and the "Read/Write Head"

The input to a TM is regarded as being contained on a *"tape"*. The tape is
divided into *"tape squares"*, so that each symbol of the input sequence occupies
one square of the tape. We assume that the tape is not restricted in length, so
that we can place an input sequence of any length we like on the tape (we assume
that any unused part of the tape consists of *blank* tape squares). For example,
the input sequence *aabbaa* may occupy part of a tape such as that depicted in
Figure 7.1.

Hitherto, our abstract machines have been assumed to *read* their input
sequences from left to right, one symbol at a time. However, a TM can move
along its tape in either direction, examining, or "reading" one tape square at a
time.

Let us now consider the output of a TM. The TM simply uses its tape for
output as well as input. On reading an input symbol from a particular tape square
in a particular state, the TM can replace the input symbol on that particular
square with any designated symbol. As a simple example, consider a tape containing a sequence of *a*s and/or *b*s, where we require as output a sequence representing
the input sequence, but with all *b*s replaced by *a*s, and *vice versa*.

An example tape is shown in Figure 7.2, as the hypothetical TM described
immediately above has completed its processing of the first three symbols (assume
that the tape is being processed from left to right).

Figure 7.1 The tape of a Turing machine, containing the string *aabbaa*.

Figure 7.2 The *read/write head* located on a square of the tape.

In order to clarify the notion of a TM "examining" one square of a tape, and placing a symbol on a tape, we say the TM possesses a *read/write head* (we sometimes simply say *R/W head*, or just *head*). The R/W head is at any time located on one of the tape squares, and is assumed to be able to "sense" which symbol of some finite alphabet occupies that square. The head also has the ability to "write" a symbol from the same finite alphabet onto a tape square. When this happens, the symbol written replaces the symbol that previously occupied the given tape square. The read/write head can be moved along the tape, one square at a time, in ways we discuss later.

7.2.2 Blank Squares

In Figure 7.2, we examined the state of the tape of an imaginary TM that replaced *b*s by *a*s (and *vice versa*), when the machine had dealt with only some of the input symbols on its tape. Let us now consider the example tape for Figure 7.2 when the machine has processed the last symbol (the *a* at the right-hand end) of the marked portion of the tape. At this point, since the machine is replacing *a*s by *b*s, and *vice versa*, the tape will be as shown in Figure 7.3.

When our machine reaches the situation shown in Figure 7.3, it will attempt to read the symbol on the next tape square to the right, which of course is blank. So that our machine can terminate its processing appropriately, we allow it to make a transition on reading a blank square. In other words, we assume that the read/write head of a TM can sense that a tape square is blank. Moreover, the TM can write a symbol onto a blank square and even "blank out" an occupied tape square.

We denote the blank symbol by an empty box (suggesting that it is represented as a blank tape square), thus:

$$\square$$

Note that in some books, the blank is represented as a 0, or a β (Greek letter *beta*).

Figure 7.3 The *read/write head* moves past the right-hand end of the marked
portion of a tape, and reaches a blank square.

7.2.3   TM "Instructions"

Like *finite state recognisers* (Chapter 4) and *pushdown recognisers* (Chapter 5), TMs can be represented as drawings, featuring *states* and labelled *arcs*. An arc in a TM is, like the arc between two states of a PDR, labelled with *three* symbols (though what the three symbols mean is different, as we will now see).

By way of illustration, Figure 7.4 shows part of a TM. Assume that the machine is in state A and that the tape and position of the read/write head are as shown on the right of Figure 7.4.

The TM represented in Figure 7.4 would carry out the activities described in Table 7.1, assuming that it starts in state A.

We now use the conceptual framework provided above as the basis of a more rigorous definition of TMs.

Figure 7.4 A fragment of a TM, along with an appropriate tape set up.

Table 7.1 A trace of the behaviour of the Turing Machine fragment from Figure 7.4.

Action	State of tape immediately following action
Read an a, write an a, move right (R), stay in state A	
Read an a, write an a, move right (R), stay in state A	
Read a b, write an a, move right (R), stay in state A	
Read a c, write a *blank*, move right (R), move to state B	

7.2.4   Turing Machines Defined

A TM has essentially the same pictorial formulation as the FSR or PDR, except that each arc of a TM is labelled with *three* instructions:

1. *an input symbol* which is either

$$\square$$

   or a symbol from a finite alphabet

2. an *output* symbol, either

$$\square$$

   or a symbol from the same alphabet as for (1),

3. a *direction designator*, specifying whether the read/write head is to move one square left (L), right (R), or not at all (N).

The TM can make a transition between states if (1) is the current tape symbol, in which case the TM replaces (1) by (2), and moves one tape square in the direction specified by (3), before moving to the state pointed at by the relevant arc.

Next, we consider a simple example TM, which is discussed in some detail. Much of the remainder of the chapter is devoted to demonstrating that TMs can perform all the tasks that can be performed by the other abstract machines we have hitherto considered, and many tasks that they could not.

7.3 The Behaviour of a TM

Figure 7.5 shows our first full TM, T_1.

T_1 processes input tapes of the form shown in Figure 7.6.

As shown in Figure 7.6, T_1 (Figure 7.5) expects the marked portion of its input tapes to consist of an a followed by zero or more blank squares, followed by either b or c. T_1's head is initially positioned at the a, and the output is a tape of the form depicted in Figure 7.7.

As Figure 7.7 shows, the b or c of the input tape (Figure 7.6) has been "erased" from its initial position – replaced by a blank – and placed in the tape square immediately to the right of the a, with T_1's head finally positioned at that square. We will call this moving of a symbol from one part of a tape to another "*shuffling*" the symbol. We can talk of *right shuffling* and *left shuffling*. T_1 does a *left shuffle* of the symbol denoted by x in Figure 7.6.

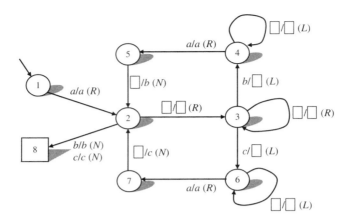

Figure 7.5 The Turing machine, T_1.

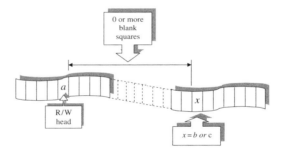

Figure 7.6 The input tape set up for T_1 of Figure 7.5.

Figure 7.7 The tape when T_1 of Figure 7.5 halts, having started on a tape as specified in Figure 7.6.

We will not trace the behaviour of T_1 in detail here. Table 7.1 shows an example of how we can trace a TM's behaviour. You can do such a trace of T_1's behaviour for yourself, if you wish. Now we will spend a short time examining the function of the states in the machine:

State 1

The *start state*. Assumes that the head is positioned at a square containing an a. If there is no a, or another symbol occupies the square, no transition is possible.

States 2–5 and 2–7

When state 2 is entered from state 1, a right move of the head has just been carried out. Since there are expected to be 0 or more blanks between the a and the b or c, the head could at this point be pointing at a b or a c (in which case the machine has nothing else to do, and moves to its halt state, leaving the head where it is). Otherwise, there is a blank to the right of the a. The machine then moves into state 3, where it "skips over" (to the right) any more blanks, until the head reaches the b or c. State 3 also "erases" the b or c before moving to state 4 (for a b) or state 6 (for a c). State 4 skips left over any blanks until the head reaches the a, at which point the head moves back to the blank square to the right of a. State 5 then ensures that the blank to the right of a is overwritten by b, as the transition back to state 2 is made (note that no movement of the head is specified by this transition). States 6 and 7 perform an analogous function to states 4 and 5, but where the symbol after the blanks is c rather than b.

State 8

This state, the *halt state*, is reached only from state 2, when the head points at a b or a c (which could be either because a b or a c was initially next to the a, or because a b or a c had been moved there by states 3–5 or states 3–7 (see above). Note that the arcs leaving states 5 and 7 could have been drawn directly to state 8 if wished.

T_1 will reach its halt state, producing an output tape as described in Figure 7.7, when it is given an *input configuration* as specified in Figure 7.6. If the input configuration is not set up correctly, then T_1 may do one of the following:

a) halt in a non-halt state, as would result from the input tape configuration shown in Figure 7.8

b) never halt, which would be the result of the input tape configuration shown in Figure 7.9.

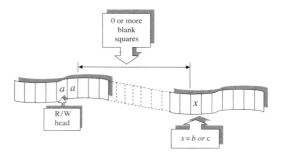

Figure 7.8 A tape that would cause problems for T_1 of Figure 7.5. There is an unexpected a to the right of the head, which would cause T_1 to get stuck in state 2.

Figure 7.9 A tape that would cause even bigger problems for T_1 of Figure 7.5. There is no b or c present, so there is nothing to stop T_1 going along the tape forever (in state 3).

You may like to satisfy yourself that (a) and (b) are true for the input configurations of Figures 7.8 and 7.9, respectively. You should also observe that a third general possibility for a Turing machine's halting situation, i.e., reaches halt state, but produces output not conforming to specification, is not possible in the case of T_1, if only blanks lie to the right of the b or c. The latter comment relates to the fact that the output specification requires a totally blank tape to the right of the head when T_1 halts.

The immediately preceding discussion indicates the need for care in describing a TM's expected *input configuration* (i.e., the marked portion of its tape, and initial location of the read/write head). If we carefully specify the input conditions, then we need not consider what happens when an input situation that does not meet our specification is presented to the machine. Of course, we should also specify the ways in which the *output configuration* (the state in which the machine halts, the contents of its tape, and the final location of its read/write head) is to represent the "result".

The remainder of this chapter will be devoted to demonstrating that for language recognition, TMs are indeed more powerful than the other abstract machines we have studied thus far.

7.4 Turing Machines as Language Recognisers

7.4.1 Regular Languages

We do not have to expend much effort to demonstrate that TMs are at least as powerful as *finite state recognisers* (FSRs). Recall from Chapter 4 that for each *regular language*, there exists a *deterministic* FSR (DFSR) to recognise that language. A DFSR is simply a TM that always moves its read/write head to the right, replacing each symbol it reads by that same symbol, on each transition.

Thus, the DFSR, $M_5{}^d$, from Chapter 4, reproduced here as Figure 7.10, becomes the TM, T_2, shown in Figure 7.11.

We have taken advantage of the TM's ability to write, as well as read, symbols, and its ability to detect the end of the input string when reaching a blank square, to ensure that T_2 has two useful features:

1. it prints T for any valid string, and F for any invalid string, and

2. it has only one halt state.

We henceforth assume that all of our TMs have only one halt state. An exercise asks you to convince yourself that this is a reasonable assumption.

An important point should be noted. T_2 prints F and goes directly to its halt state if the head is initially located on a blank square. In a sense, this reflects an

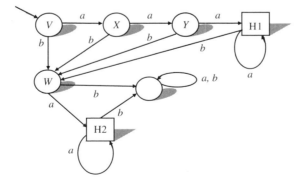

Figure 7.10 The deterministic finite state recogniser, $M_5{}^d$, of Chapter 4.

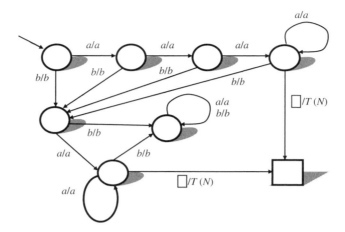

Figure 7.11 Turing machine T_2, which does the same job as DFSR $M_5{}^d$ (Figure 7.10). Since all instructions but three involve a rightward move, the move designator (R) has been omitted. Every state apart from those that write T is assumed to write F and move to the halt state if a blank square is encountered.

assumption that if the head is initially at a blank square then the tape is entirely blank. We will call this assumption the *blank tape assumption*. We will further assume that a blank tape represents the *empty string* (ε).

As ε is not in the language accepted by $M_5{}^d$, T_2 (of Figure 7.11) prints an F if its head is initially on a blank square. Of course, T_2 can never be *sure* that the tape is empty. No TM can ever establish beyond doubt that its tape is unmarked. As we know, the tape has no definite ends in either direction, i.e. the blank squares at either end of the marked portion extend indefinitely. It should therefore be clear why we have not designed T_2 so that when its head points initially at a blank square, it searches its tape to see if the input string is located somewhere else on the tape.

From the way T_2 was constructed, you will appreciate that we could do the same type of DFSR-to-TM conversion for any DFSR.

7.4.2   Context Free Languages

Recall from Chapter 5 that the abstract machine associated with the *context free languages* is the *non-deterministic pushdown recogniser* (*NPDR*). Recall also that some of the context free languages are *deterministic* (can be recognised by a *DPDR*) while some are not (can be recognised by an NPDR, but not a DPDR).

It can be shown that any DPDR can be converted into a TM. We will not consider this here, but merely point out that the basic method is for the TM to use part of its tape to represent the stack of the PDR, thus working with two parts of the tape, one representing the input string, the other representing the stack. However, one usually finds that a TM to do the job done by a DPDR is better designed from scratch. Here, we consider T_3, which is equivalent to $M_3{}^d$ of Chapter 5 (Figure 5.8), the DPDR that accepts the language

$$\{a^i b^i : i \geq 1\},$$

which we know from Chapter 6 to be *properly* context free, i.e., context free but not regular. T_3 is shown in Figure 7.12.

T_3 expects on its tape an arbitrary string of as and/or bs (using the notation of Chapter 2, a string in $\{a, b\}^*$), finally printing T if the string is in the language $\{a^i b^i : i \geq 1\}$, and F if it is not. Note that, like T_2 (Figure 7.11), T_3's read/write head begins on the leftmost symbol of the input. Moreover, T_3 embodies the same *blank tape assumption* as did T_2. T_3 uses the symbols X and Y to "tick off" the as and corresponding bs, respectively. You should trace T_2's behaviour on example sentences and non-sentences. An example of how a trace can be done is shown in Table 7.1. Now for the *non-deterministic* context free languages. Recall from Chapter 5 that some context free languages can be accepted by a *non-deterministic PDR (NPDR)*, but not a *deterministic* PDR. Later, in Chapter 11, we briefly consider the conversion of NPDRs into TMs. However, to demonstrate that TMs are more powerful than NPDRs, we see in Figure 7.13 a TM, T_4, that recognises the language of possibly empty *palindromic* strings of as and bs. Recall

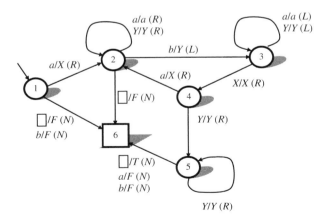

Figure 7.12 The Turing machine T_3 recognises the deterministic context free language $\{a^i b^i : i \geq 1\}$.

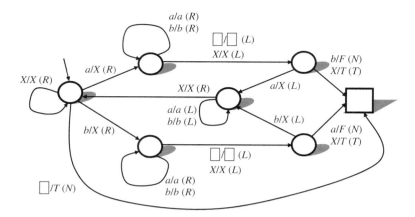

Figure 7.13 The Turing machine T_4 deterministically recognises a language of palindromic strings that requires a non-deterministic pushdown recogniser.

that in Chapter 5 such languages were shown to be non-deterministic CFLs and therefore beyond the processing power of the DPDR.

T_4 accepts only those strings in $\{a,b\}^*$ that are palindromic, and, what is more, does the job *deterministically*.

T_4 uses only one marker, X, and works from the outside in, "ticking off" matching symbols at the extreme right and left ends of the string. This time, the blank tape assumption results in a T being printed if the read/write head of T_4 starts on a blank, as the empty string is in $\{a,b\}^*$ and is palindromic. Again, you should make sure you understand how the machine works, since more complex TMs follow.

7.4.3 Turing Machines are More Powerful than PDRs

The above discussion implies that Turing machines are as powerful, in terms of language processing, as FSRs and (deterministic and non-deterministic) PDRs. This claim is justified later in this chapter. However, by using a single example we can show that TMs are actually *more* powerful than PDRs. Consider the Turing machine, T_5, shown in Figure 7.14, that processes the language

$$\{a^i b^i c^i : i \geq 1\},$$

which was demonstrated in Chapter 6 to be properly *context sensitive* (i.e. context sensitive but not context free).

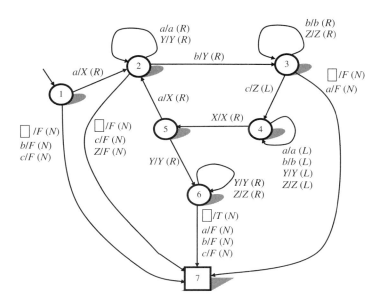

Figure 7.14 The Turing machine T_5 recognises the context sensitive language $\{a^i b^i c^i : i \geq 1\}$.

T_5 requires an input configuration of the form shown in Figure 7.15.

It reaches its halt state for valid or invalid strings. For *valid* strings (i.e. strings of *one or more as followed by the same number of bs followed by the same numbers of cs*), the output configuration is as specified in Figure 7.16.

The input sequence of Figure 7.15 is altered by T_5 to produce a tape of the form described in Figure 7.16 (the symbols X, Y and Z are used as auxiliary

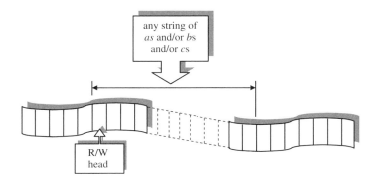

Figure 7.15 The input tape set up for T_5 of Figure 7.14.

Figure 7.16 The tape after a valid string has been processed by T_5 (Figure 7.14).

markers). Again, the output of a T indicates that the input sequence represents a sentence of the language in question.

For *invalid* strings (i.e. any string in $\{a,b,c\}^*$ which is not in $\{a^i b^i c^i : i \geq 1\}$), however, when the halt state is reached, the configuration will be as shown in Figure 7.17.

In the case of invalid strings then, the read head will be left pointing at an F, as shown in Figure 7.17.

When students design language recognition machines such as T_5, their machines sometimes omit to check for the possible occurrence of extra symbols at the end of an otherwise valid string. For the language $\{a^i b^i c^i : i \geq 1\}$, such machines will indicate that a valid string immediately followed by another string in $\{a, b, c\}^+$ is valid, when, of course, it is not.[1]

Figure 7.17 The tape after an invalid string has been processed by T_5 (Figure 7.14).

---

[1] Recall from Chapter 2 that $\{a, b, c\}^+$ is the set of all non-empty strings of as and/or bs and/or cs.

However, students are not the only guilty parties when it comes to failing to check for invalid symbols at the end of an otherwise valid input. I have encountered a Pascal compiler that simply ignores all of the source file text that occurs after an end followed by a full stop (in Pascal, the only end terminated by a full stop is the last one in the program and corresponds to a begin that announces the start of the program block). This caused severe difficulties to a novice who had accidentally put a full stop after an end halfway down their program and happened to have an extra begin near the beginning. The program compiled perfectly, but when run only carried out the first part of what was expected by the programmer! The Pascal compiler in question was accepting an invalid input string, in this case with potentially serious consequences.

7.5 Introduction to (Turing Machine) Computable Languages

We have seen that TMs are more powerful language recognisers than finite state and pushdown recognisers. Regular languages are associated with finite state recognisers in the same way as context free languages are associated with pushdown recognisers. It seems reasonable to ask, then, if there is a class of languages that are associated with TMs in a similar way. In other words, is there a class of *grammar* in the *Chomsky hierarchy*, so that for any language generated by a grammar in that class, there is some TM that can recognise that language? Conversely, for any TM that recognises a language, can we be sure that the language is in the same class? In fact, there is such a class of languages and it turns out that it is exactly that defined by the most general Chomsky classification, i.e. the general *phrase structure languages* themselves, these being those generated by the type 0, or *unrestricted*, grammars. Type 0 grammars were defined in Chapter 2 as being those with productions of the form

$x \rightarrow y, x \in (N \cup T)^+$,   i.e. a non-empty arbitrary string of terminals and

$y \in (N \cup T)^*$,   non-terminals on the left-hand side of each

production, and a possibly empty arbitrary string of

terminals and non-terminals on the right-hand sides.

Let us call the languages that can be recognised by a TM the *Turing machine computable languages*. For convenience, and with some justification, as we will see later in the book, we will refer to them as *computable languages*. We eventually discover that the computable languages include, *but are not restricted to*, all of the regular languages, all of the context free languages, and all of the context sensitive languages.

In Chapter 4, the correspondence between regular languages and finite state recognisers was specified. We saw how we can take any regular grammar and construct an FSR to accept exactly the same set of strings as that generated by the grammar. We also showed that we could take any FSR and represent it by a regular grammar that generated the same set of strings as that accepted by the FSR (we did a similar thing with respect to NPDRs and the context free languages in Chapter 5). Analogously, to show the correspondence between TMs and unrestricted languages we would have to show how to create a type 0 grammar from any TM, and *vice versa*. However, this is rather more complicated than is the case for the more restricted grammars and machines, so the result will merely be sketched here.

We first consider the context sensitive languages, as their defining grammars, the *context sensitive grammars*, cannot have the empty string on the right-hand side of productions. We then generalise our method to deal with the type 0 grammars in general. The definition of context sensitive productions is as for type 0 productions given above, with the additional restriction that the right-hand sides of context sensitive productions are not allowed to be shorter in length than the left-hand sides.

7.6 The TM as the Recogniser for the Context Sensitive Languages

The machine T_5, shown in Figure 7.14, recognises the language

$$\left\{ a^i b^i c^i : i \geq 1 \right\}.$$

In Chapter 6 we established that this language is not context free. We also saw that it is, in fact, a context sensitive language, as it is generated by the following grammar, G_{16}:

$$S \rightarrow aSBC \mid aBC$$

$$CB \rightarrow BC$$

$$aB \rightarrow ab$$

$$bB \rightarrow bb$$

$$bC \rightarrow bc$$

$$cC \rightarrow cc$$

We wish to establish that for any context sensitive language we can provide an equivalent TM. T_5 itself is not enough to convince us of this, as it was *designed* for a specific language, and not constructed through using a method for creating TMs from the productions of the corresponding grammar. We therefore require a method for constructing TMs for context sensitive grammars that we could apply to any such grammar. Here, we will consider the application of such a method to the grammar G_{16}. I then leave it to you to convince yourself that the method could be applied to any context sensitive grammar.

7.6.1 Constructing a Non-deterministic TM for Reduction Parsing of a Context Sensitive Language

The TM, T_6, that has been constructed from the productions of G_{16}, is partly depicted in Figure 7.18.

You will note that T_6 is in three sections, namely sections A, B and C, all of which have *state* 1 in common. Sections A and B are shown in full, while section C is only partly specified. The machine is far more complex than T_5 (Figure 7.14), a TM that was designed to recognise the same language. T_6 was constructed from the productions of a grammar (G_{16}), and its purpose is merely to demonstrate a general method for constructing such TMs.

The first thing to note is that T_6 is *non-deterministic*. Its overall process is to model the *reduction* method of parsing introduced in Chapter 3. Beginning with a *terminal string*, if the right-hand side of any of our productions is a substring of that string, we can replace it by the left-hand side of such a production. The process is complete if we manage to reduce the string to the start symbol S alone. For a TM, this means we finish up in a halt state with the tape that initially contained the input string now containing only S.

T_6 carries out reduction parsing using the grammar G_{16}, by effectively "scanning" the input string repeatedly from left to right, then right to left, attempting to find a sequence of reductions that will leave only the symbol S on the tape. The process is non-deterministic since many reductions may apply at any one time, and many reductions may lead to dead ends, and thus we may often have to *backtrack* to find alternative reduction sequences. As for all of the other non-deterministic machines we have studied in this book, we assume that the string is accepted if some sequence of transitions will lead to the machine's halt state (in the case of T_6, this is state 6). T_6 reaches its halt state when only S, the *start symbol* of the grammar, remains on its tape, i.e. if it finds a sequence of reduction applications that enable the input string to be reduced to S. You may like to

Figure 7.18 The *non-deterministic* Turing machine T_6 recognises the context sensitive language $\{a^i b^i c^i : i \geq 1\}$.

convince yourself that T_6's halting conditions will be met only if its input tape initially contains a string from $\{a^i b^i c^i : i \geq 1\}$.

To establish that we could produce a "T_6-like" TM from any context sensitive grammar, we briefly consider the functions of the three sections shown in the diagram of T_6, in Figure 7.18, and also the correspondence between parts of the machine and the productions of the grammar.

Section A

This section of the machine repeatedly scans the input string (state 1 scans from left to right, while state 2 scans from right to left). States 3–5 test if the symbol S alone remains on the tape (in which case the machine can halt in state 6).

Section B

This section contains a state for each production in G_{16} that has a left-hand side of the same length as its right-hand side (e.g., $CB \rightarrow BC$). For example, state 1 and state 10 replace BC by CB, i.e., carrying out the *reduction* based on the production $CB \rightarrow BC$, while state 1 and state 7 perform an analogous function for the productions $bB \rightarrow bb$ and $bC \rightarrow bc$, respectively.

Section C

This section also carries out reduction substitutions, but for productions whose right-hand sides are greater in length than their left-hand sides (in G_{16}, these are $S \rightarrow aSBC$ and $S \rightarrow aBC$). In such cases, a substitution uses up less tape than the sequence of symbols being replaced, so the extra squares of tape are replaced by blanks. Any other symbols to the right of the new blanks are then *shuffled* up one by one until the occupied part of the tape is packed (i.e., no blanks are found between any non-blank tape squares). Figure 7.18 shows only *Part* 1 of section C. We first discuss this part, before examining *Part* 2 (Figure 7.21) below.

Part 1

This section of T_6 deals with the production $S \rightarrow aBC$, i.e. "reduces" aBC to S. To illustrate, Figure 7.19 shows what happens when we are in state 11 with the tape configuration as shown.
Note that, for simplicity, Figure 7.18 shows only the loop of states (states 17, 18, 19, 15, 16) that shuffle an a leftwards so that it goes immediately to the right of the next non-blank symbol to its left. We have omitted the ten states that would be needed to shuffle any other symbols of the alphabet (i.e., b, c, S, B, C) that may be required. To consider why two states are needed to deal with each symbol, consider the part of T_6 that does the shuffling job for a, depicted in Figure 7.20.
So, states 15, 16 and 17 are common to the shuffling loop for each symbol in the alphabet. We need, say, states 20 and 21 to shuffle any bs, states 22 and 23 to shuffle any cs, states 24 and 25 to shuffle any Ss, states 26 and 27 to shuffle any Bs, and states 28 and 29 to shuffle any Cs. That is why in Figure 7.18 the state to the right of state 17 is named state 30.

Part 2

This second part of *Section* C of T_6, depicted in Figure 7.21, deals with the production $S \rightarrow aSBC$, i.e. "reduces" $aSBC$ to S. It is very similar to *Part* 1 (see Figure 7.18), that deals with $S \rightarrow aBC$, and was discussed immediately above.

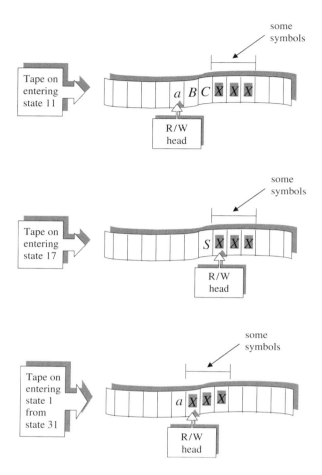

Figure 7.19 Stages in the reduction of aBC to S by the Turing machine T_6 (the state numbers refer to those in Figure 7.18).

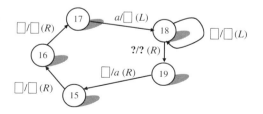

Figure 7.20 The fragment of T_6 (Figure 7.18) that "shuffles" an a leftwards over a blank portion of the tape so it is placed immediately to the right of the next non-blank tape square.

Figure 7.21 The part of T_6 that reduces $aSBC$ to S (for Part 1 of section C, see Figure 7.18).

As for *section* C *part* 1, we have only included the pair of states (state 41 and state 42) between states 40 and 38 that shuffle the a to the left (note that in this case, shuffling involves *three* blanks, as the right-hand side of $S \rightarrow aSBC$ has four symbols). As for *Part* 1, we assume that there is a similar pair of states linking states 40 and 38 for each of the other alphabet symbols (b, c, S, B, C). Thus we need ten more states (assumed to be numbered 46–57), which is why the state leaving the loop has been labelled 53. We have taken advantage of the fact that both productions dealt with by *section* C have a as the leftmost symbol of the right-hand side. This was not necessary (as our machine is non-deterministic), but was done for convenience.

7.6.2   The Generality of the Construction

To appreciate the generality of the construction that enabled us to produce T_6 from G_{16}, consider the following. The definition of context sensitive productions tells us that the left-hand side of each production does not exceed, in length, the right-hand side (i.e. productions are of the form $x \rightarrow y$, where

$|x|\leq|y|$). Therefore we can represent any context sensitive grammar by a
Turing machine with:

a *section* A part

To scan back and forth over the input between the start state and another
state, enabling the halt state to be reached only when the start symbol, S,
appears alone on the tape.

a *section* B part

That consists of a distinct path for each production in the grammar with equal
length left- and right-hand sides, each path leaving and re-entering the start
state. The path for each production performs the reduction associated with
that production, i.e., when the right-hand side of a production appears anywhere on the input tape, it can be replaced by the left-hand side of that
production.

a *section* C part

That consists of a distinct path for each production in the grammar with
left-hand side shorter in length than the right-hand side, each path leaving
and re-entering the start state. As in section B, the path for each production
performs the reduction associated with that production. However, in section C
cases the newly placed symbols occupy less tape space than the symbols
replaced, so each path has a "shuffling routine" to ensure that the tape
remains in packed form. In G_{16}, only the productions $S \rightarrow aSBC$ and
$S \rightarrow aBC$ are of section C type, and both represent reductions where the
replacement string is a single symbol. In general, of course, a context sensitive
production could have *several* symbols on the left-hand side. This represents
no real problems however, as demonstrated by the sketch of part of a TM that
does the reduction for the context sensitive production $XY \rightarrow PQR$, shown in
Figure 7.22.

As you can see from Figure 7.22, in this example we need only to skip over one tape
square to the right of the newly written XY. This is because the right-hand side of
the production $XY \rightarrow PQR$ is one symbol longer than the left. We thus replace the
PQ with XY, and the R with a blank. We then need to move over that blank to see
if there is a symbol there to be shuffled up, and so on.

In Chapter 5, the non-deterministic pushdown recognisers (NPDRs) that we
created from the productions of context free grammars recognised the language
generated by the grammar by being able to model any derivation that the
grammar could carry out. Our *non-deterministic* TMs operate by modelling
every possible sequence of reductions for the corresponding context sensitive

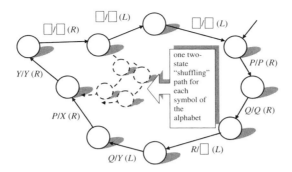

Figure 7.22 A Turing machine fragment to replace PQR by XY, representing a reduction involving the type 0 production $PQR \rightarrow XY$.

grammar. Such TMs reach their halt state only if the string on the input tape can be reduced to S. The construction sketched out above ensures that this would happen only for inputs that were sentences of the language generated by the original grammar.

Enough has been said to establish the following:

• The recogniser for the context sensitive languages is the Turing Machine. Context sensitive languages are computable languages.

You may be wondering why the statement immediately above does not say "*non-deterministic* Turing machine". In Chapter 10 we will see that any non-deterministic TM can be simulated by a deterministic TM. Hence, the non-deterministic TMs of this chapter could be replaced by equivalent deterministic TMs.

7.7 The TM as the Recogniser for the Type 0 Languages

Now, the sole difference between context sensitive productions and *unrestricted*, or *type* 0, productions is that the latter are not restricted to having right-hand sides that are greater than or equal in length to the left-hand sides. This means that, say, a production such as $WXYZ \rightarrow Pa$, which is not a valid context sensitive production, is a perfectly reasonable type 0 production. Moreover, type 0 grammars can contain ε productions (see Table 5.1). Productions such as $WXYZ \rightarrow Pa$ do not present a severe problem, as we simply add another section to our TMs to deal with these types of production, as discussed next (we will return to ε productions shortly).

7.7.1 Amending the Reduction Parsing TM to Deal with Type 0 Productions

Assuming that we already have sections A, B and C as defined for our TMs derived from context sensitive grammars (for example, Figure 7.18), we add to such a construction a new section, as follows:

Section D

Consists of a distinct path for each production in the grammar with left-hand side longer in length than the right-hand side, each path leaving and re-entering the start state. As in sections B and C, the path for each production performs the reduction associated with that production. However, in such cases the newly placed symbols occupy *more* tape space than the symbols replaced, so each path has a "shuffling routine" to ensure that all symbols in the way are each moved to the right an appropriate number of spaces. This makes room for the additional symbols required by the application of the reduction.

Figure 7.23 shows part of the section D component of a TM, in this case the part dealing with the type 0 example production $WXYZ \rightarrow Pa$.

For our example production, there are two extra symbols to insert when performing the associated *reduction*, as the right-hand side of our production – Pa – has two fewer symbols than the left-hand side – $WXYZ$. A machine of the type illustrated in Figure 7.23 therefore has to shuffle all symbols (if any) that were to the right of the Pa in question two tape squares to the right, leaving two blanks immediately to the right of Pa (which is eventually changed to WX). Then YZ can be inserted. The symbol M is used as a marker, so that the machine can detect when to stop shuffling. When the M has a blank next to it, either there are no symbols to its right, or we have shuffled the symbol next to it to the right, and so no more symbols need shuffling. You should convince yourself that a *section* D-type construction of the type shown in the example could be carried out for any production where the right-hand side is non-empty and of smaller length than the left-hand side.

7.7.2 Dealing with the Empty String

As promised above, we now consider ε-productions. In fact, we can treat these as a special case of the *section* D-type construction. We simply allow for the fact that on either side of any non-blank symbol on the tape, we can imagine there to be as many εs as we like. This means we can insert the left-hand side (i.e. after shuffling to make sufficient room, in a *section* D-type way) of a type 0 ε-production

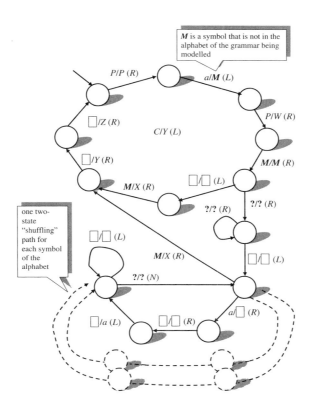

Figure 7.23 A Turing machine fragment to replace Pa by $WXYZ$, representing a reduction involving the type 0 production $WXYZ \rightarrow Pa$.

anywhere we like on the marked portion of the tape. As you may recall, in earlier chapters we noted that this was a cause of problems in parsing. Our machine should only be able to reach its halt state if it inserts the left-hand sides of ε-productions in the appropriate places, i.e. in terms of a correct reduction parse of the input string being processed. So, we complete our specification of the TM to represent any type 0 grammar by introducing the following, where necessary, to the machine sketched out in Figure 7.18.

Section E

Consists of a distinct path for each production in the grammar with ε on the right-hand side, each path leaving and re-entering the *start state*. As in sections B, C and D, the path for each production performs the reduction associated with that production. Figure 7.24 shows a sketch of a *section* E construction for the type 0 production $XYZ \rightarrow \varepsilon$.

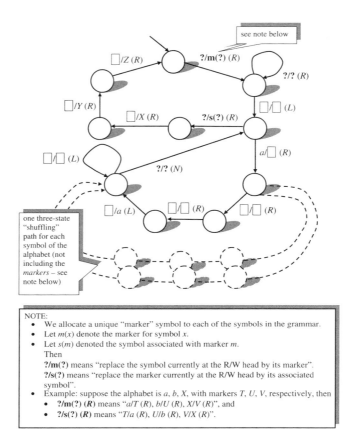

Figure 7.24 A Turing machine fragment to insert XYZ in the input tape, thus modelling the reduction based on the type 0 production $XYZ \rightarrow \varepsilon$.

The *section* E construction, as illustrated in Figure 7.24, inserts the left-hand side of an ε-production *after* any symbol on the tape. There are two problems remaining. Firstly, we must deal with an insertion *before* the leftmost symbol. Secondly, we need to cater for an insertion on a blank tape. Considering the second case first, if ε is in the language generated by the grammar, the corresponding machine should be able to reach its halt state appropriately. Our *blank tape assumption* dictates that we represent ε as the input string by setting up our machine so that its head points initially to a blank square. We can solve both problems with very similar constructions to *section* E, though there are a few differences. I leave you to sketch the construction yourself.

7.7.3   The TM as the Recogniser for all Types in the Chomsky Hierarchy

Enough has now been said to support the following claim:

- The recogniser for the phrase structure languages in general is the Turing machine. Phrase structure languages are computable languages.

(For similar reasons to those discussed above, we again refer to the "TM", not the "*non-deterministic* TM".)

We argued above that any DFSR could be converted into an equivalent TM. It was then implied that TMs could recognise any of the context free languages. However, by showing that any type 0 grammar can be represented by a TM that accepts the same language, we have also shown that the same applies to the other types of language in Chomsky's hierarchy. This is because *any production of any grammar of type* 0, *type* 1, *type* 2 *and type* 3 *is catered for in one of the sections of our construction.*

However, if we were to implement the non-deterministic TMs that are produced by our construction, we would have very inefficient parsers indeed! That is the whole point of defining restricted types of machine to deal with restricted types of language.

7.8 Decidability: A Preliminary Discussion

We close this chapter by briefly discussing one of the most important concepts related to language processing, i.e., *decidability*. In doing this, we anticipate a more thorough treatment of the topic in Chapter 11.

7.8.1   Deciding a Language

Consider the TMs T_2, T_3, T_4 and T_5 in this chapter (Figures 7.11 to 7.14, respectively). Each accepts an input tape containing a terminal string from the appropriate alphabet. If each of these machines is started with a valid input configuration, it will halt in a halt state, and as it reaches its halt state it will:

- print T if the input sequence was a sentence of the language, otherwise print F.

A language-recognising TM that conforms to the above behaviour is said to *decide* the corresponding language (because given any string of the appropriate terminals it can say "yes" if the string is a sentence *and* "no" if the string is not a sentence).

For example, consider some of the decisions made by T_5 (of Figure 7.14), that decides the language $\{a^i b^i c^i : i \geq 1\}$, as shown in Table 7.2.

The formal definition of *deciding* a language is given in Table 7.3.

Table 7.2 Various input sequences that are accepted or rejected by T_5 of Figure 7.14.

Input string	T_5 input and output
aabbcc (valid string)	initial configuration: `a a b b c c` (R/W head on first a)
	final configuration: `X X Y Y Z Z T` (R/W head on T)
aaabbcc (invalid string – too many *a*s)	initial configuration: `a a a b b c c` (R/W head on first a)
	final configuration: `X X X Y Y F Z` (R/W head on F)
caaaa (invalid string – starts with *c*)	initial configuration: `c a a a a a` (R/W head on c)
	final configuration: `F a a a a a` (R/W head on F)

Table 7.3 Deciding a language.

Given some language L, with alphabet A, a TM *decides* that language if for any string, x, such that $x \in A^*$:

if $x \in L$

    the machine indicates its acceptance,

and

if $x \notin L$

    the machine indicates its rejection

of x.

A language for which there is a TM that *decides* that language is called a *decidable* language.

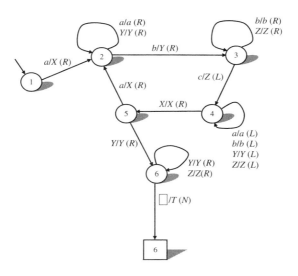

Figure 7.25 The Turing machine T_7 (based on T_5 of Figure 7.14) also recognises the context sensitive language $\{a^i b^i c^i : i \geq 1\}$. Unlike T_5, T_7 does not print F for invalid strings.

7.8.2 Accepting a Language

Suppose we amend T_5 (of Figure 7.14) to produce the machine T_7, depicted in Figure 7.25.

We have simply removed from T_5 all arcs entering the halt state which result in writing the symbol F, and removed the instructions that write an F from the arc between states 6 and 7. Like T_5, T_7 will now reach its halt state and write a T for valid input strings. However, unlike T_5, for invalid strings T_7 will eventually stop in one of its non-halt states, and make no F indication at all. For example, given the input *aaabbcc* (the second example used for T_5 in Table 7.2), T_7 would eventually come to a stop in state 2, being unable to make any move from that state in the way that T_5 could.

A machine that can output T for each and every valid string of a language (and never outputs T for any *in*valid strings) is said to *accept* that language. Compare this carefully with the definition of *deciding*, from Table 7.3. Clearly, a machine that decides a language also accepts it, but the converse is not necessarily true. T_5, T_6 and T_7 all *accept* the language $\{a^i b^i c^i : i \geq 1\}$, but only T_5 *decides* it.

The formal definition of accepting a language is given in Table 7.4.

Table 7.4 Accepting a language.

Given some language L, with alphabet A, a TM *accepts* that language if for any string, x, such that $x \in A^*$:

if $x \in L$

     the machine indicates its acceptance of x.

A language for which there is a TM that *accepts* that language is called an *acceptable* language.

7.9 End of Part One ...

In this chapter, we have encountered *computable languages, decidable languages,* and now *acceptable languages*. From the way in which the construction of TMs from type 0 grammars was developed, it can be seen that such TMs would *accept*, rather than decide, the language generated by the grammar from which the machine was constructed. In other words, *computable languages are acceptable languages*. However, we have also seen that some computable languages are decidable (since TMs T_2, T_3, T_4 and T_5 all *decide* their respective languages). A question addressed in Chapter 11 asks whether all computable languages are also *decidable*. We find that in general they are not, though the more restricted languages of the Chomsky hierarchy are.

In Chapters 2 and 6, we saw the type 0 grammar, G_4, which generates what we call the *multiplication language*:

$$\left\{ a^i b^j c^{i \times j} : i, j \geq 1 \right\}.$$

In discussing this language in Chapters 2 and 5, it was remarked that we could regard phrase structure grammars as *computational*, as well as *linguistic*, devices. Subsequently, in Chapter 6, we saw that the multiplication language could not be generated by a context free grammar, which tells us that a pushdown recogniser is insufficiently powerful to accept the language. From the present chapter we know that we could define a TM, from the productions of G_4, that *would* accept this language. This seems to imply that the TM has a certain computational ability that is worthy of clearer definition and exploration. Such an enterprise has played a central part in computer science, and it is a major aim of the second part of this book to discuss this. However, this does not mean that we have finished with languages. Far from it. It can in fact be shown that the computation performed by any TM can be modelled by a type 0 grammar. The proof of this is complex and we will not consider it in this book, but the result is very important, since again it establishes a general equivalence between a class of languages and a class of abstract machines. More than this, since the TM is the acceptor for the type 0 languages and thus for all the subordinate language types, such a result establishes a general equivalence between TMs and the *entire* Chomsky hierarchy.

The TM is the first machine we have seen in this book that is capable of producing a string of symbols (on its tape) as a *result*, when given a string of symbols as its *input*. This means that the TM can carry out forms of computation that can provide richer answers than "yes" or "no". The second part of this book investigates the nature of the computational power of the TM. In the process of this, we discover that the TM is more powerful than any digital computer. In fact, the TM is the most powerful computational model we can define. Eventually, our investigations bring us full circle and enable us to precisely specify the relationship between *formal languages, computable languages* and *abstract machines*. In doing so, we are able to answer a question that was asked at the end of Chapter 6: are there *formal languages* that are a proper *superset* of the *type* 0 languages?

EXERCISES

For exercises marked "†",solutions,partial solutions,or hints to get you started appear in "Solutions to Selected Exercises"at the end of the book.

7.1.[†] Sketch out a general argument to demonstrate that any Turing machine needs only one halt state, and also that the single halt state requires no outgoing arcs.

7.2. Design Turing machines to decide the following languages:

(a) $\{a^{2i+1}c^jb^{2i+1} : i,j \geq 1\}$

(b) $\{xx : x \in \{a,b\}^*\}$

(c) $\{a^ib^jc^{i+j} : i,j \geq 1\}$

(d) $\{a^ib^jc^{i-j} : i,j \geq 1\}$

(e) $\{a^ib^jc^{i \times j} : i,j \geq 1\}$

(f) $\{a^ib^jc^{i \ div \ j} : i,j \geq 1\}$ (*div* is *integer division*)

Hint: in all cases, first design the machine so that it deals only with acceptable strings, then add additional arcs so that it deals appropriately with non-acceptable strings.

7.3. The following is a recursive definition of "well-formed parenthesis expressions" (WPEs):

$()$ is a WPE

if X is a WPE, then so is (X)

if X and Y are WPEs, then so is XY

(a) Define a TM to decide the language of all WPEs.

(b) The language of all WPEs is context free. Show this by defining a context free grammar to generate it.

Hint: the recursive definition above converts almost directly into a CFG.

Part 2
Machines and Computation

8
Finite State Transducers

8.1 Overview

In this chapter, we consider the types of computation that can be carried out by *Finite State Transducers* (*FST*s), which are the *Finite State Recognisers* of Chapter 4, with output.

We see that FSTs can do:

- various limited tasks involving memory of input sequences

- *addition* and *subtraction* of arbitrary length numbers

- restricted (constant) *multiplication* and *modular division*.

We then find that FSTs:

- cannot do multiplication of arbitrary length numbers

- are not a very useful model of real computers (despite some superficial similarities with them).

8.2 Finite State Transducers

A *finite state transducer* (*FST*) is a *deterministic finite state recogniser* (*DFSR*) with the ability to output symbols. Alternatively, it can be viewed as a *Turing machine* (*TM*) with two tapes (one for input, one for output) that

A.P. Parkes, *A Concise Introduction to Languages and Machines*,
DOI: 10.1007/978-1-84800-121-3_8, © Springer-Verlag London Limited 2008

- always moves its *head* to the *right*

- cannot recognise *blank squares* on its *tape*.

As the machine makes a transition between two states (according to the input symbol it is reading, of course), it outputs one symbol. In addition to performing recognition tasks, such a machine can also carry out computations. The purpose of this chapter is to investigate the scope of the FST's computational abilities. We shall see that we need a more powerful type of machine to carry out all of the computations we usually expect machines (i.e. computers) to be able to do. As we will see, this more powerful machine is the one we encountered in Chapter 7. It is the TM.

8.3 Finite State Transducers and Language Recognition

First of all, you should appreciate that the FSRs used for *regular language* recognition in Chapter 4 are simply a special case of FSTs. We could have defined FSRs so that on each transition they produce an output symbol, and we could have defined these output symbols to be significant in some way. In our first example machine, we assume that, when the input is exhausted, the *last* symbol the machine outputs is a 1, then the input string is *accepted*, and if the last symbol output is a 0, then the string is *rejected*. Figure 8.1 shows such a machine.

The FST in Figure 8.1 accepts the language $\{a^i b^j : i, j \geq 1\}$. It indicates acceptance or rejection of an input string by the last symbol it outputs when

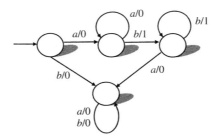

Figure 8.1 A finite state transducer that recognises the regular language $\{a^i b^j : i \geq 1\}$. The last symbol it outputs indicates if it accepts (1) or rejects (0) the input string.

Table 8.1 Example input strings, and corresponding output strings, for the finite state transducer of Figure 8.1. The last symbol output indicates the machine's decision.

Input string	Output string	ACCEPT or REJECT
aaab	0001	ACCEPT
aa	00	REJECT
aabba	00110	REJECT
aabbb	00111	ACCEPT

the last symbol of the input has been read ($1 = accept, 0 = reject$). Some examples of the machine's output are shown in Table 8.1. The significant symbol of the output (the last) is underlined in each case.

As you can see, the machine in Figure 8.1 has no state designated as a *halt state* or *accept state*. This is because the *output* is now used to tell us whether or not a string is accepted.

8.4 Finite State Transducers and Memory

An interesting application of FSTs is as "memory" devices. For example, FST_1, in Figure 8.2, processes inputs that are strings of binary digits. As output, the machine produces strings of binary digits, representing the input string "shifted" two digits to the right.

You may think that the names of FST_1's states look strange. Of course, the state names do not affect the operation of the machine, but they have been chosen for symbolic reasons, which should become apparent as the chapter continues.

The next machine, FST_2, does a similar task, but with a right shift of *three* digits. FST_2 is shown in Figure 8.3.

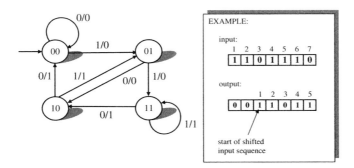

Figure 8.2 FST_1, a binary two-digit "shift" machine.

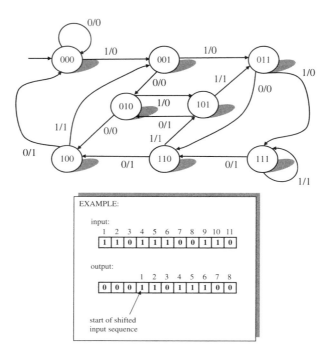

Figure 8.3 FST_2, a binary three-digit "shift" machine.

Now consider FST_3, shown in Figure 8.4, that does a "shift two to the right" operation for inputs consisting of strings in $\{0, 1, 2\}^+$ (remember from Chapter 2 that A^+ denotes the set of all non-empty strings that can be taken from the alphabet A).

Machines such as FST_{1-3} are sometimes called "memory machines", as they perform a simple memory task. Imagine that a "clock" governs the processing of the input, a concept we will return to later in the book. At each "tick" of the clock, the next symbol is read and the appropriate symbol is output. Then:

FST_1

stores a symbol of a two-symbol alphabet for two "ticks" and then outputs it,

FST_2

stores a symbol of a two-symbol alphabet for *three* "ticks" and then outputs it,

and,

FST_3

stores a symbol of a *three*-symbol alphabet for *two* "ticks" and then outputs it.

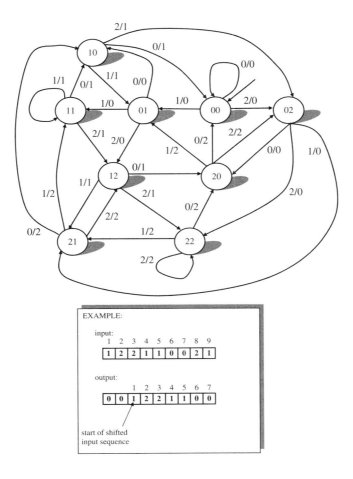

Figure 8.4 FST_3, a "trinary" two-digit "shift" machine.

You should be able to see that we could design memory machines for an alphabet of any number, s, of symbols, and for any number, t, of clock ticks, *providing that we know in advance the values of s and t*. You should also be able to see that we need to increase the number of states as the values of s and t increase.

In fact, and this is where the names of the states in our three machines are significant, there is a simple numeric relationship between the number of symbols, s, in the alphabet, the number of ticks, t, that each input symbol has to be "remembered" for, and the minimum number of states required to do the job. There are s^t (numerically speaking) distinct sequences of length t when the alphabet has s symbols, and the only way the FST can remember which of the s^t sequences it has just read is to have a state for each one, so the FST needs a minimum of s^t states. As you can see, the state names in the machines in this section represent, for each machine, the s^t sequences of length t we can have for the s-symbol alphabet in question.

So,

FST_1 where $t = 2$ and $s = 2$, has $s^t = 2^2 = 4$ states,
FST_2 where $t = 3$ and $s = 2$, has $s^t = 2^3 = 8$ states,
and
FST_3 where $t = 2$ and $s = 3$, has $s^t = 3^2 = 9$ states.

8.5 Finite State Transducers and Computation

In this section, we shall see that FSTs possess reasonably sophisticated computational abilities, so long as we are fairly creative in defining how the input and/or output sequences are to be interpreted. As an example of such "creativity", we will first look in a new way at the tasks carried out by our machines FST_{1-3} (Figures 8.2 to 8.4), in the preceding section.

It should be pointed out that there is no trickery in the way we are going to interpret the input and output of our machines. We are simply defining the type of input expected by a particular machine and the form that the output will take. This, of course, is a crucial part of problem solving, whether we do it with formal machines or real programs.

8.5.1   Simple Multiplication

Beginning with FST_1, the machine that shifts a binary sequence two digits to the right, let us consider that the sequence of 1s and 0s presented to the machine is first subjected to the following (using 1011 as an example):

1. The string 00 is appended to the rightmost end (giving 101100, for the example).

2. The resulting sequence is interpreted as a binary number *in reverse* (for the example we would then have 001101, which is 13 in *decimal* (base 10).

The output sequence is then also interpreted as a binary number in reverse.

For the example sequence 101100, FST_1 would produce as output 001011, which in reverse (110100) and interpreted as a binary number is decimal 52.

You can convince yourselves that under the conditions defined above, FST_1 performs *multiplication by* 4.

For FST_2, we apply similar conditions to the input, except that for condition (1) – see above – we place *three* 0s at the rightmost end of the input sequence. FST_2 then performs *multiplication by* 8.

For FST_3, the same conditions apply as those for FST_1, except that the input and output number are interpreted as being in *base* 3, in which case FST_3 performs *multiplication by* 9. Thus, given the input 121100 (i.e. 001121, or 43 decimal), FST_3 would produce as output 001211 (i.e. 112100), which, according to our specification, represents 387 (i.e. 43 × 9).

We have seen that FSTs can carry out certain restricted multiplication tasks. Later in this chapter we will see that there are limitations to the abilities of FSTs with respect to multiplication in general.

8.5.2   Addition and Subtraction

FSTs are capable of adding or subtracting binary numbers of any size. However, before we see how this is done, we need to define the form of our input and output sequences.

The input

When *we* add or subtract two numbers (assuming that the numbers are too long for us to mentally perform the calculation), we usually work from right to left (least significant digits first). It seems reasonable, then, in view of the need to deal with any "carry" situations that may arise, that our machines can do likewise. In the previous section we came across the notion of the input being in reverse. Clearly, this method can also be used here, and will then represent the machine processing the pairs of corresponding digits in the way that we usually do.

Apart from certain minor problems that will be addressed shortly, the remaining major problem is that we have *two* numbers to add or subtract, whereas our machines accept only one sequence as input. Our solution to this problem is to present the two binary numbers (to be added or subtracted) as *one sequence*, in which the digits of one of the numbers are interleaved with the digits of the other. This, of course, means that the numbers must be of the same length. However, we can ensure that this is the case by placing sufficient 0s at the leftmost end of the shorter of the two numbers before we reverse them and present them to our machines.

Table 8.2 Data preparation for binary addition and subtraction FSTs.

To prepare two binary numbers for input to the addition or subtraction FSTs:

1 Ensure numbers are the same length by padding the shorter number with leading 0s.

1a (For addition only) Place an extra leading 0 on each number, to accommodate possible final carry.

2 Reverse both numbers.

3 Make one sequence by interleaving the corresponding digits of each reversed number.

(Note: for subtraction FST, begin with second number.)

Table 8.2 summarises the preparation of input for the FSTs for addition and subtraction which we will define shortly.

We now turn to the output.

The output

For our addition and subtraction FSTs to model our behaviour, they must process the two numbers by examining pairs of corresponding digits, in least to most significant digit order of the original numbers. However, as an FST can only read one digit at a time, it will not be able to output a 0 or 1 representing a digit of the answer until it has examined the second digit of each pair of corresponding digits. Moreover, our FSTs must output a symbol for each input symbol encountered. For the first digit in each corresponding pair, then, our FSTs will output the symbol "*". Note that there is no special significance to this symbol, it is simply being used as a "marker".

Thus, given input of the form indicated by the diagram that accompanies step 3 in Table 8.2, our FSTs will produce output as represented in Figure 8.5.

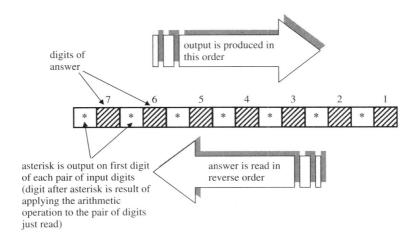

Figure 8.5 Specification of the output (in this case, for a seven-digit number) for the arithmetic finite state transducers considered in this chapter.

Figure 8.6 How we construct the answer from the output specified in Figure 8.5.

The output shown in Figure 8.5 represents the digits of the answer *in reverse*, with each digit preceded by *. The *s are not part of the answer, of course, and so we read the answer as specified in Figure 8.6.

The machines

We are now ready to present the two machines, FST_4, the binary adding FST (Figure 8.7), and FST_5, the binary subtractor (Figure 8.8). Each machine is accompanied by an example showing a particular input and the resulting output. I leave it to you to convince yourself that the machines operate as claimed.

It is important to note that for both machines there is no limit to the length of the input numbers. Pairs of binary numbers of any length could be processed. It is clear, therefore, that the FST is capable of adding and subtracting arbitrary length binary numbers.

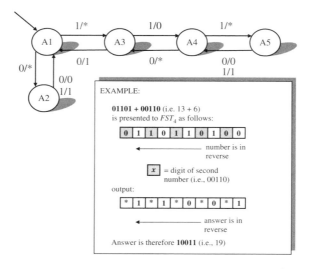

Figure 8.7 A binary adder finite state transducer, FST_4.

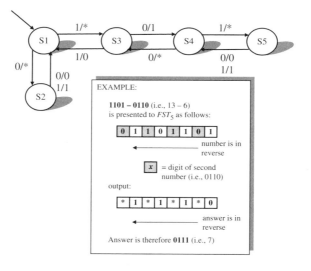

Figure 8.8 A binary subtractor finite state transducer, FST_5. If the result of subtracting the numbers would be negative, the result is undefined.

8.5.3   Simple Division and Modular Arithmetic

Certain restricted forms of division are also within the capabilities of the FST. For example, FST_6, in Figure 8.9, performs integer division of a binary number by 3, returning the result as a binary number. In some programming languages, integer division is represented by the infix operator *div*, so FST_6 actually does x *div* 3, where x is the input number.[1] However, FST_6 does not require the input sequence to be in reverse. FST_6 carries out integer division from the *most* significant end of the number, as we usually do. The machine also outputs the result in the correct order. Figure 8.9 includes an example input sequence and its corresponding output sequence.

We can quite easily use FST_6 as the basis of a machine that computes the quotient *and* the remainder, by making the following observations about states of FST_6:

$D0$ represents a remainder of 0 (i.e. no carry),
$D1$ represents a remainder of 1 (i.e. a carry of 1),
$D2$ represents a remainder of 2 (i.e. a carry of 2).

Now, the remainder resulting from a division by 3 requires a maximum of *two* binary digits, as it can be 0, 1, or 10 (i.e., 0, 1 or 2). We could make sufficient space for the remainder, by placing "***" (three asterisks) after the input sequence.

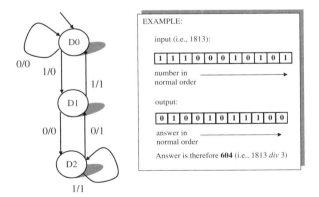

Figure 8.9 A binary "*div* 3" finite state transducer, FST_6.

---

[1] An infix operator is written in between its operands. The arithmetic *addition* operator, "+", is thus an example of an infix operator.

I leave it to you to decide how to amend the machine to ensure that the quotient is followed by the binary representation of the remainder. The first of the three asterisks should be left unchanged by the machine, to provide a separator between quotient and remainder.

We could also amend our machine to provide only the remainder. For this we would ensure that all the outputs of FST_6 were changed to "*", and then carry out the amendments discussed in the preceding paragraph. Then our machine would carry out an operation similar to that called *mod* in programming languages, except it would be restricted to *mod* 3.

8.6 The Limitations of the Finite State Transducer

We have seen that FSTs can be defined to do restricted forms of multiplication and division and also that FSTs are capable of unrestricted addition and subtraction. You will note that the preceding sentence does not include the word "binary". This is because we could define adders and subtractors for any number base (including base 10), since there are only a finite number of "carry" cases for addition and subtraction in *any* given number system. Similarly, we can define FSTs to do multiplication or division by any fixed multiplicand or divisor, in any number base. However, the binary machine is much simpler than would be the corresponding FST for, say, operations on base 10 numbers. Compare the complexity of FST_1 (Figure 8.2) and FST_3 (Figure 8.4), for example. I will leave it to you to consider the design of FSTs for number systems from base three upwards.

In fact, our intuition tells us that the binary number system is sufficient; after all, we deal every day with a machine that performs *all* of its operations using representations that are essentially binary numbers. We never use that as a basis for arguing that the machine is insufficiently powerful! We return to the issue of the universality of binary representations several times in subsequent chapters.

For now, our purpose is to establish that FSTs are insufficiently powerful to model all of the computational tasks that we may wish to carry out. We have seen that the FST can perform unrestricted addition and subtraction. The question arises: can we design an FST to do unrestricted multiplication? We shall eventually demonstrate that the answer to this question is "no". This is obviously a critical limitation, since the ability to perform multiplication is surely one of the basic requirements of useful computing systems. First, however, we show that there are *some* multiplication operations we *can* perform with FSTs.

8.6.1 Restricted FST Multiplication

Earlier in this chapter, we saw that FSTs are capable of multiplying a base n number by some constant c, where c is a power of n (for example, FST_1 was used to multiply a binary – base 2 – number by 4, i.e. 2^2). Suppose we wish to multiply a number, say a binary number, by a number that cannot be expressed as an integral power of 2, say 3. One way to achieve such tasks is to effectively "hard-wire" the multiplicand 3 into the machine, in a similar way that the divisor 3 was represented in the div 3 machine, FST_6 (Figure 8.9).

FST_7, shown in Figure 8.10, multiplies an arbitrary length binary number by 3.

To see how FST_7 works, consider its operation in terms of writing down a binary number three times and adding the three copies together. Figure 8.11 represents the *manual* version of the example shown in Figure 8.10.

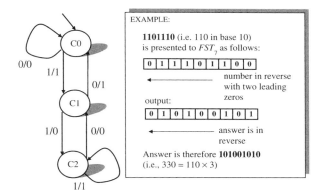

Figure 8.10 A binary " $\times$ 3" finite state transducer, FST_7.

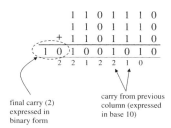

Figure 8.11 Multiplying a binary number by 3 by adding the number three times. The calculation is the same as that represented by the example in Figure 8.10. See also Table 8.3.

Table 8.3 Possible carry situations, expressed as base 10 numbers, for multiplication by 3 by adding together three copies of a binary number.

		Column digit			
		1		0	
Carry	0	1	1	0	0
	1	0	2	1	0
	2	1	2	0	1
		result	new carry	result	new carry

In adding together three identical binary numbers, beginning from the rightmost column, we can either have three 1s (i.e. $1 + 1 + 1$) $= 11$, i.e. 1 carry 1, or three zeros (0 carry 0) i.e. a carry of 1 or 0. If there is a carry of 0, we continue as in the previous sentence. Otherwise, we have a carry of 1 and the next column could be three 1s (i.e. $1 + 1 + 1 + 1$ – adding in the carry) so we write down a 0 and carry 2 (decimal). If the *next* column is three 0s, we have $0 + 0 + 0 + 10 = 10$, i.e. 0 carry 1 – back to carry 1, which has been discussed already. Otherwise, we have three 1s and a carry of 10, i.e. $1 + 1 + 1 + 10 = 101 = 1$ carry 2 – back to carry 2, which, again, we have already considered. The whole situation is summarised in Table 8.3.

The states in Figure 8.10 were labelled $C0$–2 to represent the three carry situations found in Table 8.3. From Table 8.3, you should be able to see why only three states are required for FST_7. Moreover, you should also convince yourself that we could not define an FST of fewer than three states to carry out the same task.

A similar method can be used to create a machine, FST_8 that multiplies a binary number by 5. The machine is shown in Figure 8.12 and, as you can see, requires five states.

Finally, we consider, in Figure 8.13, the rather unusual FST_9, which multiplies a base 7 number by 4 (!).

I leave it to you to convince yourself that the above machines, FST_{7-9}, operate as claimed.

The above multipliers, FST_{7-9}, reflect a construction suggesting that a machine to multiply a number in any base by a constant c, needs no more states than c (i.e. one for each of the possible base 10 carry values). As we saw above:

FST_7, where $c = 3$, had 3 states,
FST_8, where $c = 5$, had 5 states,
FST_9, where $c = 4$, had 4 states.

Similar observations can be made with respect to the number of states required by a constant divisor machine (consider FST_6 in Figure 8.9).

However, you may like to consider what happens to our construction when the numbers to be multiplied are in bases greater than 10.

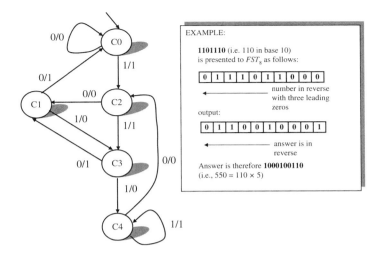

Figure 8.12 FST_8 multiplies any binary number by 5.

Figure 8.13 FST_9 multiplies any base 7 number by 4.

8.6.2   FSTs and Unlimited Multiplication

In this section we see that FSTs are incapable of multiplication of arbitrary length numbers. The argument used is essentially an informal application of the *repeat state theorem* introduced in Chapter 6. Again, the method of "proof" used is *reductio ad absurdum*.

We can assume, for reasons that will be discussed in later chapters, and as hinted at earlier in this chapter, that it is sufficient to focus on the task of multiplying arbitrary length *binary* numbers. Our assumption is that we have an FST, M, that can multiply arbitrary length pairs of binary numbers and print out the result. Now, we can assume that M has a fixed number of states. Say it has k states. We ask M to multiply the number 2^k by itself:

2^k is a 1 followed by k 0s, in binary number notation.
$2^k \times 2^k$ (i.e., 2^{2k}) is a 1 followed by $2k$ 0s
(example: $2^3 = 1000$, $2^3 \times 2^3 = 2^6 = 1000000$, i.e. $8 \times 8 = 64$).

Suppose M reads the corresponding digits of its input numbers simultaneously. M has only k states and it must use at least one of them to process the input digits. However, after it has processed the input digits it still has to print out the rest of the 0s in the answer. Let's assume that M uses only one of its states to do the multiplication (an extremely conservative estimate!) and print out the 1 followed by the first k 0s of the answer (one digit for each of the k pairs of input digits it encounters). It then has only $k - 1$ states left with which to print out the remaining k 0s of the answer. M must therefore *loop* when it prints out the rest of the 0s. It is printing out the 0s without any further input of digits, so there is no way of controlling the number of 0s that M prints. The only way to make an FST generate a fixed number, k, of symbols without input is to have a sequence of $k + 1$ states to do it. This notion was discussed in the context of finite state recognisers in Chapter 4 and formed the basis of the *repeat state theorem* of Chapter 6.

We are therefore forced to reject our initial assumption: the machine M cannot exist. Multiplication of arbitrary length binary numbers is beyond the power of the FST. By extension of this argument, multiplication of arbitrary length numbers in *any* number base is also beyond its power.

8.7 FSTs as Unsuitable Models for Real Computers

It is unfortunate that FSTs are not powerful enough to perform all of the operations that we expect computers to do. As you can probably guess from considering the decision programs that represent FSRs (Chapter 4), FSTs also lend themselves to conversion into very simple programs. However, at the very least, we expect any reasonable model of computation to be able to deal with arbitrary multiplication and division. As we have seen, this basic requirement is beyond the capabilities of FSTs.

You may think you detect a flaw in an argument that assumes that real computers can perform operations on arbitrarily large data. Most of us who have ever written any non-trivial programs have experienced having to deal with the fact that the computer we use is limit*ed*, not limit*less*, in terms of its storage and memory capacity. Since a computer is, in terms of available storage at any given point in time, finite, it could be regarded as a (huge) FST. As such, for any given computer, there is, even for basic operations such as multiplication, a limit to the size of the numbers that can be dealt with. In this sense, then, a computer is an FST.

It turns out, however, that treating a digital computer as if it were an FST is not, in general, a useful position to adopt, either theoretically or practically.[2] Let us look at the practical issues first. When a deterministic FST enters the same state (say q) twice during a computation, it is looping. Moreover, once the machine reaches state q, repeating the input sequence that took it from q back to q (perhaps via several other states) will result in the same behaviour being exhibited again. Once again, this was the basis of the repeat state theorem for FSRs in Chapter 6.

Suppose a program is performing some calculations, subsequently to print some output at the terminal, and receives no output from external sources once it has begun (i.e., it operates on data stored internally). The *state* of the computer running the program is its entire internal configuration, including the contents of its memory, its registers, its variables, its program counter, and so on. Viewed in this way, if such a state of the computer is ever repeated during the execution of the program, the program will repeat the whole cycle of events that led from the first to the second occurrence of that configuration over and over again, *ad infinitum*. If this happens we have obviously made a (fairly common) mistake in our coding, resulting in what is often called an *infinite loop*.

It would be useful if, before we marketed our programs, we could test them to make sure that such infinite loops could never occur. Let us suppose that we have some way of examining the internal state of the machine each time it changes, and some way of deciding that we had reached a state that had occurred before.

---

[2] This is not altogether true. The FST is used, for example, to model aspects of computer networks (see Tanenbaum, 1998, in the "Further Reading" section). I am referring here to its ability to model arbitrary computations.

In theory, this is possible because, as stated above, a computer at a given point in time is a finite state machine.

Suppose that our computer possesses 32 Kilobytes of total storage, including data, registers, primary memory, and so on. Each individual bit of the 32 K can be either 1 or 0. Thus we have a total of 262 144 bits, each of which can be either 1 or 0. This means that, at any point our machine is in one of $2^{262\ 144}$ states. $2^{262\ 144}$ is about $10^{79\ 000}$. This is an *incredibly* large number. In Chapter 10 we come across a *really* fast (over three million million instructions per second) hypothetical machine that would take 10^{80} *years* (a 1 followed by *eighty* 0s) to solve a problem involving passing through 10^{100} states. Our computer may thus have to go through a huge number of internal states before one of these states is repeated. Waiting for even a tiny computer, like our 32 K model here, to repeat a state because it happens to be an FST is thus a rather silly thing to do!

In the remainder of this book, we study the deceptively simple abstract machine that was introduced in Chapter 7, i.e. the *Turing machine (TM)*. The TM is essentially an FST with the ability to move left and right along a single input *tape*. This means that the TM can overwrite symbols on its tape with other symbols, and can thus take action on the basis of symbols placed there earlier in its activities. In subsequent chapters, we find that this simple machine is actually *more* powerful than any digital computer. In fact, in some senses the Turing machine is the most powerful computational device of all. However, we also discover that there is something about the problem of whether or not a certain computation will eventually stop, as discussed in terms of the FST above, that puts its general solution beyond *any* computational device.

EXERCISES

For exercises marked "†", solutions, partial solutions, or hints to get you started appear in "Solutions to Selected Exercises" at the end of the book.

8.1. FSTs can be used to represent Boolean logic circuits. Design FSTs to perform "bitwise" manipulation of their binary number inputs. For machines needing two input values, assume an input set up similar to that defined for FST_4 and FST_5, the addition and subtraction machines of Figures 8.7 and 8.8, respectively. The operations you may wish to consider are:

and

or[†]

not

nand

nor

8.2[†]. Design an FST to convert a binary number into an *octal* (base 8) number.

Hint: each sequence of three digits of the binary number represents one digit (0–7) of the corresponding octal number.

9

Turing Machines as Computers

9.1 Overview

In this chapter, we begin our investigation into the computational power of *Turing machines* (*TMs*), that we first encountered in Chapter 7.

We find that TMs are capable of performing computational tasks that *finite state transducers* cannot do, such as

- multiplication, and

- division

of arbitrary length binary numbers.

We also see a sketch of how TMs could perform any of the operations that computers can do, such as:

- logical operations

- memory accessing operations, and

- control operations.

9.2 Turing Machines and Computation

In Chapter 7, *Turing machines* (*TMs*) for language recognition were introduced. We saw that the TM is the recogniser for all of the languages in the *Chomsky hierarchy*. Chapter 8 demonstrated that the *finite state transducer* (*FST*), which

A.P. Parkes, *A Concise Introduction to Languages and Machines*,
DOI: 10.1007/978-1-84800-121-3_9, © Springer-Verlag London Limited 2008

is essentially a TM that can move only to the right on its *tape*, is capable of some fairly sophisticated computational tasks, including certain restricted types of multiplication and division. We also observed that any given digital computer is essentially finite in terms of storage capacity at any moment in time. All this seems to suggest that viewing the computer as an FST might be a useful perspective to take, especially since an FST can be expressed as a very simple program. However, we saw that the huge number of internal states possible in even a small computer meant that notions typically applied to FSTs, such as the *repeat state theorem*, were of limited practical use.

In this chapter, we see that, in terms of *computation*, the Turing machine is a much better counterpart of the computer than is the FST. Simple as it is, and resulting as it does from addition of a few limited facilities to the FST, the TM actually exceeds the computational power of any single computer, of any size. In this sense, the TM is an abstraction over all computers, and the theoretical limitations to the power of the TM are often the *practical* limitations to the digital computer, not only at the present time, but also at any time in the future. However, we shall descend from these heady philosophical heights until the next chapter. For now, we concentrate on demonstrating that the TM is capable of the basic computational tasks we expect from any *real* computer.

9.3 Turing Machines and Arbitrary Binary Multiplication

Chapter 8 established that FSTs are not capable of multiplication of arbitrary length numbers, in any number base. In this section, we sketch out a TM to perform multiplication of two binary numbers of any length. I will leave it to you to realise that similar TMs could be defined for any number base (cf. the treatment of different number bases in Chapter 8). However, here we choose the binary number system, for obvious reasons.

9.3.1   Some Basic TM Operations

A side effect of defining a TM for multiplication is that we will appreciate a notion actually formulated by Turing himself in conjunction with his machines, and one which occurs in some form in various programming paradigms. Essentially, we shall build up our binary multiplier by creating a "library" of TM "subroutines". These subroutines, like subroutines in typical programming languages ("procedures" and "functions", in Pascal), represent operations that we wish to carry out many times and thus do not want to create over and over again. However, as there are no straightforward mechanisms for parameter passing in TMs, our subroutines are more analogous to *macros*, i.e. we have to imagine that the "sub-machine" that constitutes the macro is actually inserted into the appropriate place in the overall machine. This means that we have to "call" our sub-machine by ensuring that the "calling" machine configures the tape and R/W head appropriately.

Some of the operations required by our multiplier Turing machine are now specified. Our multiplication routine will work by repeated addition, based on the common "shift and add" long multiplication method (we will examine this in more detail in the next section). Clearly, then, we need an *ADD* routine. Our *ADD* routine will make use of two sub-machines, *WRITE-A-ZERO* and *WRITE-A-ONE*, as specified in Figure 9.1.

Referring to Figure 9.1, recall from Chapter 7 that we use "?/?" to mean "replace any *non-blank* symbol of the alphabet for the machine by the same symbol".

Our *ADD* routine adds two arbitrary length binary numbers, adding pairs of digits in least to most significant order (as we usually do). The machine assumes that the two numbers are initially represented on a tape as specified in Figure 9.2.

So, *ADD* expects on its tape a "*" followed by the first number, followed by a "*" followed by the second number. Its head is initially located on the least significant digit of the second number. *ADD* will not be concerned with what is to the left of the relevant part of the input tape, but will assume a blank immediately to the right of the second number. The left asterisk will be used by *ADD* to indicate the current carry situation during the computation ("*" = no carry, C = carry). The machine will write the result of the addition to the left of the currently occupied part of the input tape (and thus assumes that the first

Figure 9.1 Turing machine fragments "*WRITE-A-ZERO*" and "*WRITE-A-ONE*", as used by TMs later in the chapter.

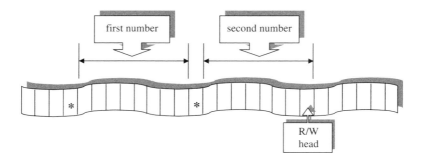

Figure 9.2 The input tape set up for the *ADD* Turing machine.

blank found to the left of the left "*" indicates the extreme right of the unoccupied left-hand side of the tape).

The description of *ADD* will be supported by an example input tape, and we will examine the effects of each major part of *ADD* on the contents of this tape. The example input tape we will use is shown in Figure 9.3.

9.3.2 The "ADD" TM

Figure 9.4 shows part one of the *ADD* TM.

Part one of *ADD*, shown in Figure 9.4, deals with the adding of corresponding digits of the two numbers until the digits of one or both of the numbers have been used up. Note that *X* and *Y* are used as markers for 0 and 1, respectively, so that *ADD* can detect how far along each number it has reached, at any stage, and thus avoid processing the same digit more than once. When the final digit of one or

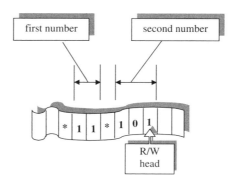

Figure 9.3 An example input configuration for the *ADD* TM. This represents the binary sum $11 + 101$ (i.e., $3 + 5$ in base 10).

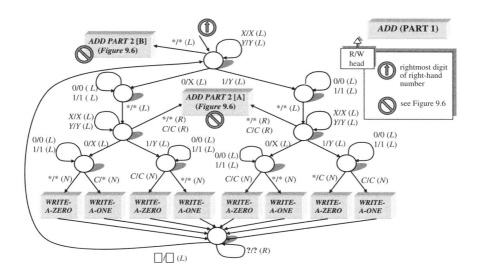

Figure 9.4 The ADD Turing machine (part 1). ADD adds together two binary numbers, beginning with a tape as specified in Figure 9.2. The answer is written to the left of the occupied portion of the initial tape.

both of the numbers has been reached, part two of ADD (shown in Figure 9.6) takes over. At the time we are ready to enter ADD part two, our example tape is configured as shown in Figure 9.5.

As can be seen in Figure 9.6, there are two entry points A and B, to ADD part two.

Entry point A

This is entered from ADD (part 1) when there are no digits remaining in the left-hand number, but a digit was read from the right number. ADD has just overwritten either a 0 (by X) or a 1 in the right-hand number (there must have been at least one more digit in the right number than in the left number or we would not have reached entry point A). Thus, this digit (represented by the leftmost X or Y in the right number) is set back to its original value, so that it can be processed as an extra digit by the sub-machine $TIDY$-UP, the definition of which is left as an exercise.

Entry point B

This is entered from ADD (part 1) when there are no digits remaining in the right-hand number. This case is simpler than that for *entry point A*, as no extra digits have been marked off (ADD always looks at the digit in the *right* number first, before it attempts to find the corresponding digit in the left number).

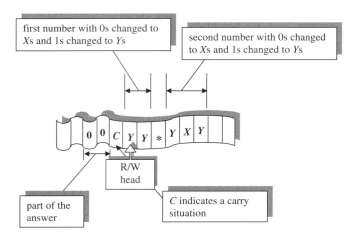

Figure 9.5 The output from ADD part 1 (Figure 9.4) for the example input of Figure 9.3. This is as the tape would be on entry to ADD part 2 (Figure 9.6). As there are fewer digits in the left-hand number, entry will be to ADD part 2 [A].

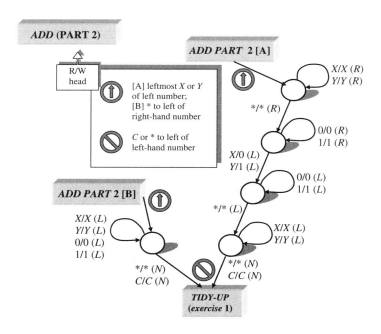

Figure 9.6 The ADD Turing machine (part 2; for part 1 see Figure 9.4). This part of ADD deals with any difference in the length of the two input numbers.

Entry points A and B both ensure that the head is located at the square of the tape representing the carry situation. Remember that this is the square immediately to the left of the left number, and is occupied with the symbol "*" for no carry and C for carry.

Our example tape (Figure 9.3) would have resulted in entering add at *entry point A*, as the end of the right-hand number was reached first. Immediately before the machine enters the component labelled *TIDY-UP*, which you are invited to complete as an exercise, our example tape from Figure 9.3 would have the configuration shown in Figure 9.7.

ADD should now complete its computation, the result being the binary number at the left of the occupied portion of the tape. For our example tape, Figure 9.8 shows an appropriate outcome from *TIDY-UP*, and thus an appropriate representation of the result of *ADD*, given the input tape of Figure 9.3.

9.3.3 The "MULT" TM

We are now ready to sketch out the description of a TM, *MULT*, to multiply two arbitrary length binary numbers. As mentioned earlier, our machine uses the "shift and add" method. For example, suppose we wish to compute 1110×101. This would be carried out as shown in Figure 9.9.

An example of the same method being employed for the multiplication of two base 10 numbers may provide a more familiar illustration. Figure 9.10 shows the method being used for the base 10 computation 1345×263.

Figure 9.7 The output from *ADD* part 2 (Figure 9.6) for the example input of Figure 9.5. This is as the tape would be on entry to *TIDY-UP*, which you are asked to construct in exercise 1 at the end of this chapter.

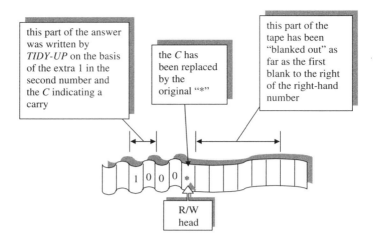

Figure 9.8 The expected output from *TIDY-UP* after processing the tape of Figure 9.7.

Figure 9.9 Multiplying two binary numbers using the "shift and add" method.

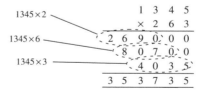

Figure 9.10 Multiplying two decimal numbers using the "shift and add" method.

The "shift and add" process is simpler for binary numbers, since the same rule applies for placing zeros at the end of each stage, but we only ever have to multiply at intermediate stages by either 1 or 0. Multiplying by 1 involves simply copying the number out exactly, while multiplying by zero can, of course, be ignored. An algorithm to do "shift and add" binary multiplication is given in Table 9.1.

Table 9.1 The "shift and add" method for multiplying two binary numbers.

$\{n_1$ and n_2 are the two numbers to be multiplied and d_i represents the ith digit from the right of $n_2\}$

begin

if $d_1 = 1$ then
      $ans := n_1$
else
      $ans := 0$
endif
	P
	A
	R
	T
	
	1

for each digit d_i, where $i \geq 2$, in n_2 do
    if $d_i \neq 0$ then
        $ans := ans + (n_1$ with $i - 1$ 0 s appended to its right)
    endif
endfor
	P
	A
	R
	T
	
	2

end

Our *MULT* machine will be based on the "shift and add" algorithm shown in Table 9.1. The detailed implementation will be left to the exercises. The *MULT* machine will expect its input to be configured exactly as that for the *ADD* TM, as specified in Figure 9.2.

An example *input configuration* for *MULT* is shown in Figure 9.11. Note that we have decided that the read/write head should be positioned initially on the rightmost digit of the right-hand number, this number being called n_2 in the algorithm of Table 9.1.

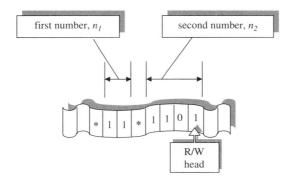

Figure 9.11 An example input configuration for the *MULT* TM.
This represents the binary calculation 11 × 1101
(i.e., 3 × 13 in base 10).

We will now sketch out the actions to be carried out by $MULT$, in terms of the algorithm of Table 9.1. The first part of the algorithm is the "if" statement (shown as "part 1" in Table 9.1).

The head is currently over the first (rightmost) digit of n_2, which in the algorithm is called d_1. If d_1 is a 0, $MULT$ initialises the answer to 0, otherwise the answer is initialised to n_1. $MULT$ will use the area to the left of n_1 as the "work area" of its tape, and leave the answer there when it terminates, separated by a single blank from the "*" preceding n_1. "$ans := n_1$" thus means "copy n_1 into the answer area". For this we need a copy routine, which we will call $COPY\text{-}L$. $COPY\text{-}L$ is defined in Figure 9.12.

$COPY\text{-}L$ needs little explanation. The symbols X and Y are again used as temporary markers for 0 and 1, respectively. The machine leaves the copied sequence exactly as it finds it. Clearly, the machine is designed to expect an alphabet of (at least) $\{*, X, Y, 0, 1\}$, which is due to the requirements of the multiplier.[1] The machine could easily be amended to deal with other alphabets.

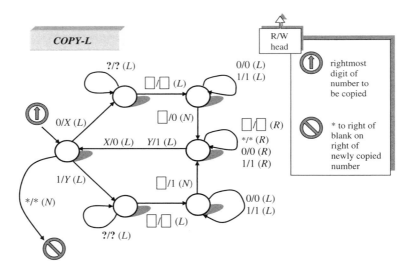

Figure 9.12 The $COPY\text{-}L$ Turing machine. Copies a binary number preceded by "*" from the position of the read/write head to the left end of the marked portion of tape. Leaves a blank square between copied sequence and existing occupied tape.

---

[1] Since $MULT$ will use ADD, the sub-machines we define for $MULT$ may need to take account of symbols used by ADD, such as C (the carry indicator), and so on.

It would also be a simple matter to create a corresponding *COPY-R* ("copy right") machine, if required.

If $d_1 = 0$, then *MULT* simply writes a 0 in the answer area. In our example here, $d_1 = 1$, so *MULT* must move its head to the rightmost digit of n_1, then enter *COPY-L*. At this point, *MULT* could also place a "*" at the left-hand end of the newly written answer sequence.

The example tape would then be as shown in Figure 9.13.

MULT will now deal with the next part of the algorithm, i.e., the "for loop" (part two of Table 9.1).

Having processed the rightmost digit, d_1, of n_2, as specified above, *MULT* now turns its attention to the remaining digits of n_2. We have seen examples of the type of processing involved here earlier in this book. TMs use auxiliary symbols to "tick off" each symbol (i.e. temporarily replace it by another symbol) so that they can return to that symbol, thus being able to locate the next in the sequence, and so on. For the algorithm above, *MULT* must examine all of the digits, d_i, of n_2 (for $i \geq 2$). If a d_i is 0, *MULT* simply proceeds to the next digit (if there is one). Otherwise d_i is a 1, so *MULT* must write a 0 at the left-hand end of its tape for each digit in n_2 to the right of d_i (this corresponds to us placing zeros at the end of a number in the "shift and add" method). Then it must do a *COPY-L* to copy n_1 immediately to the left of the leftmost 0 just written. If *MULT* then places a "*" to the left of the copied n_1, and moves the head right until it reaches a blank, the tape is now set up so that our *ADD* TM can take over. Via *ADD*, *MULT* is now carrying out the addition statement in the algorithm (Table 9.1, part two).

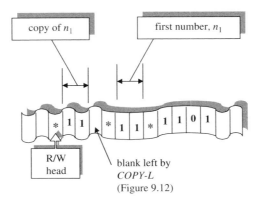

Figure 9.13 The state of the input tape from Figure 9.11, following *MULT*'s execution of the statement $ans := n_1$ (Table 9.1). The first number, n_1, has been copied to the left of n_1.

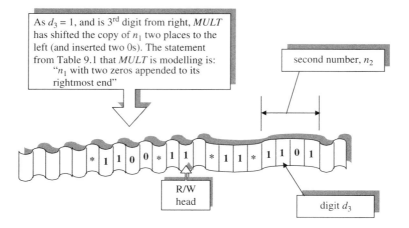

Figure 9.14 The state of the input tape from Figure 9.11 immediately before *ADD* is used by *MULT* to add 1100 to 11. *MULT* is processing digit d_3 of n_2.

For our example tape, $d_2 = 0$, but d_3, the leftmost digit of n_2, is 1. *MULT* must therefore place two zeros to the left of the answer field, then copy n_1 to the left of those zeros, meaning that the tape presented to *ADD* would be as shown in Figure 9.14.

ADD was designed so that it erases its tape from the right of the answer *as far as the first blank to the right*. This means that *ADD* would not erase *MULT*'s input numbers, since a blank separates the input to *MULT* from the answer. *MULT* would then need to prepare the tape for the next digit of n_2 (d_4 in our example) to be processed. This could involve simply "shuffling" the answer sequence up so that only one blank separates it from *MULT*'s input sequence. Our example tape would now be as displayed in Figure 9.15 (subject to some slight repositioning of the "*").

MULT would now deal with d_4 of n_2 in an analogous way to its treatment of d_3, described above. This time, *three* zeros would be needed, so the tape on input to *ADD* would be as shown in Figure 9.16.

Figure 9.17 shows a suitably tidied up form of the final result .

Sufficient detail has been presented to demonstrate that such a *MULT* machine could be implemented. The exercises ask you to do this. It is clear that such a *MULT machine* could process binary numbers of any length. We established in the preceding chapter that multiplication of arbitrary length numbers is beyond the capabilities of the *FST*. We now know, therefore, that the TM possesses greater computational power than does the FST.

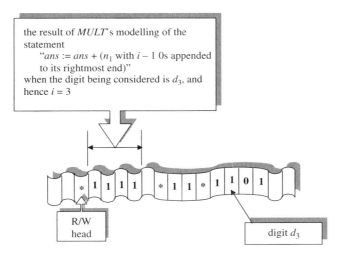

Figure 9.15 The state of the input tape from Figure 9.11 immediately after *ADD* has performed $1100 + 11$, representing the partial application of the "shift and add" multiplication procedure (see Figure 9.14 for the tape set up immediately prior to this).

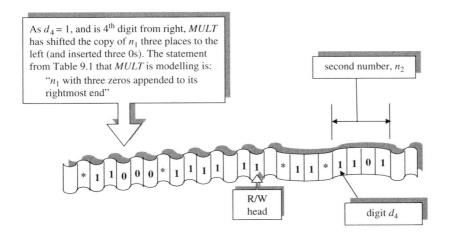

Figure 9.16 The state of the input tape from Figure 9.11 immediately before *ADD* is used by *MULT* to add 11000 to 1111. *MULT* is processing digit d_4 of n_2.

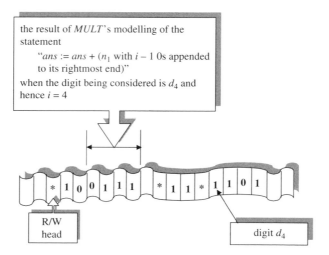

Figure 9.17 The state of the input tape from Figure 9.11 immediately after *ADD* has performed 11000 + 1111, representing the partial application of the "shift and add" multiplication procedure (see Figure 9.16 for the tape set up immediately prior to this). This is also the final output from *MULT*, representing 1101 × 11 = 100111, i.e. 13 × 3 = 39.

9.4 Turing Machines and Arbitrary Integer Division

As we can do multiplication by repeated addition, so we can do division by repeated subtraction. We repeatedly subtract one number from the other until we obtain zero or a negative result. We keep a count of how many subtractions have been carried out, and this count is our answer. In this section, we sketch a TM that does arbitrary integer division. The actual construction is left as an exercise.

Our machine, *DIV*, will perform division by repeated subtraction. We therefore require a *SUBTRACT* sub-machine for our *DIV* machine (as we needed an *ADD* machine for *MULT*).

9.4.1   The "SUBTRACT" TM

It turns out that we can do subtraction in terms of *addition*, and you may be familiar with the approach considered here. For example, 8 − 5 is really the same as saying $8 + (-5)$. If we have a way of representing the negation of the second number, we can then simply add that negation to the first number. The usefulness of this method for *binary* numbers relates to the fact that the negation of a number is simple to achieve, using a representation called the *twos complement* of a binary number. The process of obtaining the twos complement of a binary number is described in Figure 9.18.

So, to compute $x - y$, for binary numbers x and y, we compute the twos complement of y and add it to the first number. Referring to the example of $8 - 5$ mentioned above, we can represent this in binary as $1000 - 101$. The twos

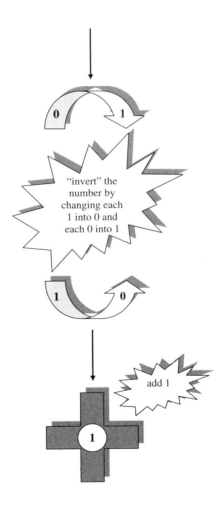

Figure 9.18 Obtaining the "twos complement" of a binary number.

complement of 101 is 011 (the inverse is 010, and $010 + 1 = 011$). $1000 + 011 = 1011$. Now, this result appears to be 11 (eleven) in base 10. However, after the addition we ignore the leftmost digit of the result, and thus our answer is 011 in binary (i.e. 3 in base 10), which is the correct result.

If you have not come across the method before, you should convince yourself that twos complement arithmetic works, and also discover for yourself why the leftmost digit of the result is ignored.

In terms of the development of the *SUBTRACT* machine, the subtraction method we have just described can be done in terms of the *ADD* TM described earlier in the chapter. We can even use *ADD* in the sub-machine that computes the twos complement, since once we have inverted a binary number, we then need to *add* 1 to the result. A TM, *TWOS-COMP*, which embodies such an approach, is described in Figure 9.19.

As you can see from Figure 9.19, *TWOS-COMP* makes use of the TM *INVERT*, which is defined in Figure 9.20.

You can study the behaviour of *TWOS-COMP* for yourself, if you wish.

We could set up our *SUBTRACT* TM so that it uses the same input format as that used for *ADD*, and shown in Figure 9.21.

We assume that the "second number" of Figure 9.21 is to be subtracted from the "first number".

The exercises ask you to complete the *SUBTRACT* machine. This simply involves using the *ADD* machine to add the twos complement of one number to the other number. The main problem, as for the *MULT* machine, concerns the

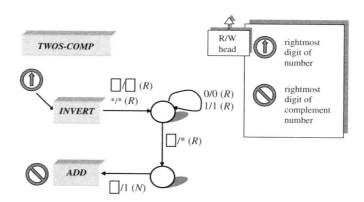

Figure 9.19 A Turing machine, *TWOS-COMP*, to compute the twos complement of a binary number (the *INVERT* machine is in Figure 9.20, the *ADD* machine in Figures 9.4 and 9.6).

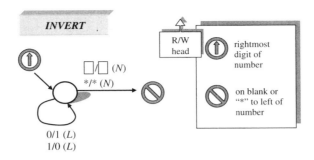

Figure 9.20 A Turing machine, *INVERT*, that is used by *TWOS-COMP* (Figure 9.19). It inverts each digit of a binary number.

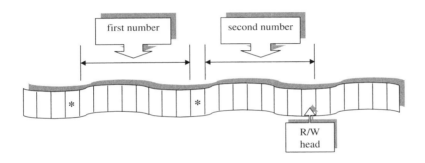

Figure 9.21 The input tape set up for the *SUBTRACT* Turing machine.

need to shuffle the various intermediate sequences around to ensure that essential information is not corrupted; but it is not too difficult to achieve this.

9.4.2 The "DIV" TM

We now sketch a TM for performing integer division by repeated subtraction. Again, its implementation is left as an exercise. An algorithm to carry out the task is specified in Table 9.2.

As for the *MULT* and *SUBTRACT* TMs, the main difficulty in implementing the *DIV* TM is in ensuring that the sub-machines are "called" properly and do not accidentally access inappropriate parts of the marked tapes. This needs careful consideration of where each sub-machine expects asterisks or blanks as delimiters, and which parts of the tape a sub-machine erases after completing its computation. One way is to use different areas of the tape (separated by one or more

Table 9.2. The "repeated subtraction" method for division, using the Turing machines *SUBTRACT* and *ADD*.

{the calculation performed is $n_1 \div n_2$}

begin

    count := 0

    temp := n_1

    while $n_2 \leq$ temp do

        temp := $SUBTRACT$(temp, n_2)

        count := ADD(count, 1)

    endwhile

    {*count* is the result}

end

> $T(x, y)$ is a functional representation of the statement: "apply the TM, *T*, to the tape appropriately configured with the numbers *x* and *y*"

blanks) for different purposes. For example, you might represent the *count* variable from the algorithm in one place, the original n_1 and n_2 in another place. Sometimes, you may have to *shuffle* up one marked area to make room for the result from another marked area. However, as the main operations that increase the tape squares used are copying and adding, your machine can always be sure to create enough room between marked areas to accommodate any results. For example, if we add two binary numbers we know that the length of the result cannot exceed the length of the longer of the two numbers by more than one digit. All this shuffling can be quite tricky and it is more important to convince yourself that it can be done, rather than to actually do it.

You will notice that the division algorithm in Table 9.2 includes the $\leq$ test. As yet we have not designed a TM to perform such logical comparisons. Therefore, in the next section we develop a generalised *COMPARE* TM. *COMPARE* may also be of use to the *SUBTRACT* machine (to detect if the number being subtracted is larger than the number from which it is being subtracted).

9.5 Logical Operations

The generalised comparison TM, *COMPARE*, tells us whether two arbitrary length binary numbers are equal, or whether one number is greater than, or less than the first number. *COMPARE*'s input configuration will follow that for *ADD*, *MULT* and *SUBTRACT*, and is shown in Figure 9.22.

COMPARE will write the symbol =, <, or > in place of the "*" between the two numbers in Figure 9.22 according to whether the "second number" is equal to, less than, or greater than the "first number", respectively. Thus, for example, given the example input tape configuration shown in Figure 9.23, *COMPARE* will produce an output configuration as shown in Figure 9.24.

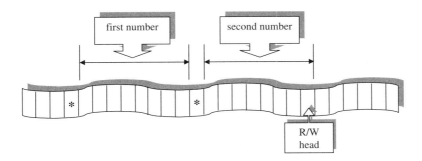

Figure 9.22 The input tape set up for the *COMPARE* Turing machine (Figures 9.25 and 9.26), that tells us the relationship between the first and second number.

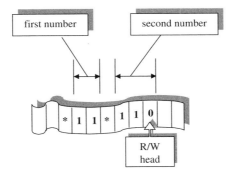

Figure 9.23 An example input configuration for the *COMPARE* TM (Figures 9.25 and 9.26). This represents the question "*what is the relationship between* 110 *and* 11?"

Figure 9.24 The output from the *COMPARE* TM (Figures 9.25 and 9.26) for the input specified in Figure 9.23. Given the original input, the output represents the statement "110 is *greater than* 11".

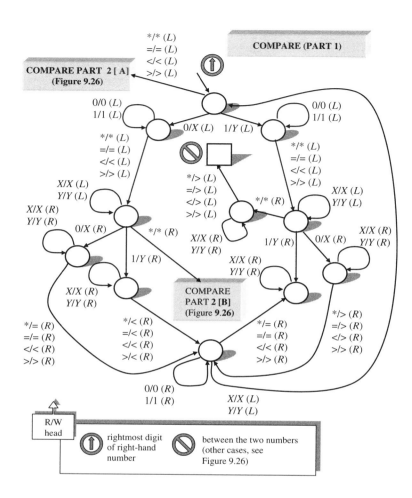

Figure 9.25 The *COMPARE* Turing machine (part 1). Does a "bitwise" comparison of two binary numbers, beginning with a tape as specified in Figure 9.22. Answer is a symbol ("<", ">", or "=") written between the two numbers.

The *COMPARE* TM is shown in Figures 9.25 (part 1) and 9.26 (part 2). Part one deals with the comparison until all of the digits of one number have been processed.

When all of the digits of one of the numbers have been processed, part 2 of *COMPARE* (Figure 9.26) takes over.

I leave it to you to follow the behaviour of *COMPARE* on some examples and to convince yourself that it operates as claimed. As part of the *DIV* machine (see

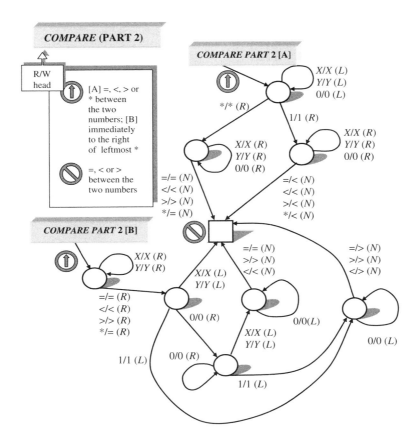

Figure 9.26 The *COMPARE* Turing machine part 2 (for part 1 see Figure 9.25). This part of *COMPARE* deals with any difference in the length of the two input numbers.

the exercises), you could use *COMPARE* to carry out the test to determine when the repeated subtractions should stop.

9.6 TMs and the Simulation of Computer Operations

TMs are extremely primitive machines, with a one-dimensional data storage device. It thus may seem strange to claim, as at the beginning of this chapter, that TMs are actually as powerful as, and in some senses *more powerful* than, *any* digital computer can be. To provide evidence for this claim, we have seen TMs that do multiplication and division of arbitrary length binary numbers. Such operations are beyond the capabilities of any machine with a fixed amount of storage, such as any real computer at a given point in time. We have also seen how logical operations such as comparison of (again, arbitrarily large) numbers can be carried out by our TMs.

In this chapter, we considered only the binary number system. You should not see this as a limitation, however, as your experience with computers tells you that ultimately, everything that is dealt with by the computer, including the most complex program and the most convoluted data structure, is represented as binary codes. Even the syntactic rules (i.e. the *grammars*) of our most extensive programming languages are also ultimately represented (in the compiler) as binary codes, as is the compiler itself.

In the following chapter, we take our first look at *Turing's thesis*, concerning the computational power of TMs. Before this, we consider more evidence supporting the claim that TMs are as powerful as any computers by briefly mapping out how many internal functions performed by, and many representations contained within, the digital computer, can be modelled by a TM. This is not an exhaustive map, but a sketch of the forms that the TM implementation of computer operations could take. I leave it to you to convince yourself of the general applicability of TMs to the modelling of any other operations found in the modern computer.

Tables 9.3 to 9.6 show examples for each specified internal operation. For each example, we suggest a TM that can model that operation (some of these

Table 9.3 Turing machine simulation of arithmetic operations.

Example operations	TM implementation	Computer limitations
Division	DIV	Size of operands/size of result
Multiplication	$MULT$	,,
Addition	ADD	,,
Subtraction	$SUBTRACT$	,,

Power (x^y)	result := 1 {write a 1 in the result area of tape}
	 while $y \neq 0$ {$COMPARE(y, 0)$ in another area of the tape}
	   result := result $\times$ x {$MULT$(result, x) in another area of tape}
	   $y := y - 1$ {$SUBTRACT(y, 1)$ in another area of tape}
	 endwhile
		,,

Table 9.4 Turing machine simulation of logical operations.

Example operations	TM implementation	Computer limitations
$<, >, =$	*COMPARE*	Length of operands
$\leq, \geq, \neq$	*Example*: "$\neq$" (Turing machine *NOT-EQ*)	

bitwise *NOT*   *INVERT*   Length of operand

are from earlier in the chapter). In each case, certain limitations are given that would apply to any digital computer, at a given moment in time, but do not apply to TMs.

Table 9.3 suggests how TMs can model arithmetic operations. In Table 9.4, logical operations are considered. Memory and array representation and accessing feature in Table 9.5, while Table 9.6 briefly considers the TM simulation of program execution (a subject which is given much more attention in Chapters 10 and 11).

It is worth focusing briefly on memory and array operations (Table 9.5), as these did not feature earlier in this chapter. In the examples given in Table 9.5 we have used memory structures where each element was a single binary number. This is essentially the memory organisation of the modern digital computer. I leave it to you to appreciate how complex data items (such as records, etc.) could also be incorporated into the scheme.

Careful consideration of the preceding discussion will reveal that we seem to be approaching a claim that Turing machines are capable of performing any of the tasks that real computers can do. Indeed, the remarks in the third columns of Tables 9.3 to 9.6 suggest that computers are essentially limited compared with Turing machines. These observations lead us on to *Turing's thesis*, which forms a major part of the next chapter.

Table 9.5 Turing machine simulation of memory/array structure and access.

Example operations	TM implementation	Computer limitations
One-dimensional array	*index* ⟷ | * | ⟷ *item₁* | * | ⟷ *item₂...* ⟷ {*index* is the number of the item to be accessed} 1. To the left of *index*, write a 0 (to represent a *count* variable initialised to 0) 2. *ADD*(*count*, 1) 3. *COMPARE*(*count*, *index*). if they are equal, proceed right to the first "*" to the right of *index*. If a blank is encountered, HALT, otherwise copy the digits between that "*" and the next "*" (or blank) somewhere to the left of *index*, then go back rightwards along the tape changing all *X*s back to *s, then HALT else proceed right to the first "*" to the right of *index*, change it to *X*, and go to 2.	Item length/address length/array size
Indirect addressing	As above, but in step 3 overwrite *index* with copied number and go to step 1.	"
Multi-dimensional arrays	*EXAMPLE*: 3 × 2 × 2 three-dimensional array i ⟷ | * | ⟷ j ⟷ | * | ⟷ k ⟷ | * | 1,1,1 – 1,2,2 | **I** | 2,1,1 – 2,2,2 | **I** | 3,1,1 – 3,2,2 | **I** | 2,1,1 | **K** | 2,1,2 | **J** | 2,2,1 | **K** | 2,2,2 *i, j,k*: indexes of item to be retrieved (i.e., item *k* of item *j* of item *i*) *K*: marker for increment of *k* index *J*: marker to set *k* item number to 1 and increment *j* item number *I*: marker to set both *k* and *j* item number to 1 and increment *i* item number	As above plus permissible number of dimensions

Table 9.6 Turing machine simulation of program execution.

Example operations	TM implementation	Computer limitations
Running a program	Use binary numbers to represent codes for operations such as *ADD*, *MULT*, memory access and so on.	Size of program
	Set up data and memory appropriately on tape. Use a portion of the tape as data, and a portion to represent sequences of operations (i.e. the *program*).	Amount of data Number of operations

EXERCISES

For exercises marked "†", solutions, partial solutions, or hints to get you started appear in "Solutions to Selected Exercises" at the end of the book.

9.1. Define the TM *TIDY-UP* for the *ADD* machine, described above.

9.2. Complete the definitions of the TMs *MULT* (using the *ADD* TM), *SUBTRACT* and *DIV* (perhaps making use of *COMPARE*).

9.3. Design a TM that accesses elements of a two-dimensional array, following the specification given in Table 9.5.

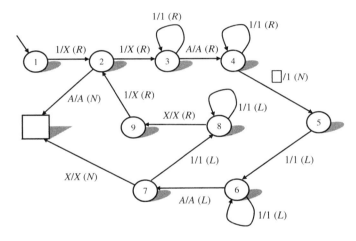

Figure 9.27 What function does this Turing machine compute?

Figure 9.28 The input configuration for the Turing machine in Figure 9.27.

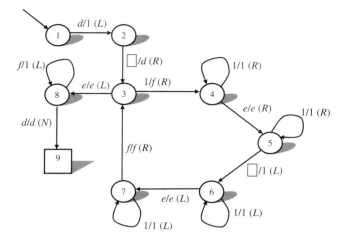

Figure 9.29 A Turing machine that does language recognition and computation.

Figure 9.30 The input configuration for the Turing machine in Figure 9.29.

9.4.[†] What *function* is computed by the TM in Figure 9.27, assuming that its input tapes are configured as specified in Figure 9.28?

In Figure 9.28, y is an arbitrary string of the form 1^n, $n \geq 1$, representing the integer n, followed by a marker, A. The output is assumed to be the integer represented by m, where m is the number of 1 s to the right of the A when the machine halts.

9.5. The Turing machine, T, is as depicted in Figure 9.29.

The initial tape is of the form shown in Figure 9.30.

Referring to Figure 9.30, note that:

d and e are alphabet symbols;

x is a (possibly empty) sequence of 1 s.

a)[†] Briefly describe the relationship between the tape on entering the loop of states 3–7, and the tape when all iterations through the loop have been completed.

b)[†] If the arc between state 3 and state 8 labelled "e/e (L)" was altered to "$e/1$ (L)", what function, expressed in terms of x, would the amended machine compute, assuming that $|x|$ (i.e., the number of 1 s on the initial tape) represents an integer, and that the result is the number of 1 s to the right of d when the machine halts?

c) Delete state 1 from T, and by amending arcs as necessary, draw a machine, T_a, which accepts the language $\{d1^n e1^n : n \geq 0\}$. State 2 of T should become the start state of T_a and state 9 of T its halt state. Additional alphabet symbols may be required, but you should introduce no additional arcs.

10

Turing's Thesis and the Universality of the Turing Machine

10.1 Overview

In this chapter, we investigate *Turing's* thesis, which essentially says that the Turing machine (TM) is the most powerful computational device of all. We find evidence to support the thesis by considering a special TM called the *Universal* Turing machine (*UTM*).

UTM is a TM that models the behaviour of any other TM, M, when presented with a binary code representing:

- M's input

- M itself.

As defined in this chapter, *UTM* is a TM with three tapes (and three associated read/write heads).

This leads us on to consider multiple-tape (*k-tape*) TMs and the *non-deterministic* TMs, introduced as the recogniser for the type 0 languages in Chapter 7. We discover that:

- any non-deterministic TM can be represented by a deterministic, 4-tape TM,

and then that

- any k-tape TM can be modelled by a standard, deterministic, single-tape TM.

A.P. Parkes, *A Concise Introduction to Languages and Machines*,
DOI: 10.1007/978-1-84800-121-3_10, © Springer-Verlag London Limited 2008

We conclude with the observation that a significant outcome of the above, with respect to Part 1 of the book, is that the standard TM is the recogniser for the type 0 languages.

10.2 Turing's Thesis

We begin by continuing a discussion of the power of the TM that began in the preceding chapter. This leads to what is known as *Turing's thesis*. What you have to appreciate about a "thesis" is that it is not a "theorem". A *theorem* is a statement that can be proved to be true by a step-by-step argument that follows valid logical rules, just as the *repeat state theorem* for finite state recognisers was "proved" in Chapter 6. A *thesis*, on the other hand, is something that somebody asserts, for some reason, as being true. In other words, anybody can state that any assertion is a thesis, but it serves little use as a thesis if someone else immediately disproves it.

However, Turing's thesis, which was initially posited in the 1930s, has yet to be disproved. Not only has it not been *dis*proved, but related work carried out by other mathematicians has led to essentially the same conclusions as those reached by Turing himself.[1] For example, an American mathematician, Alonzo Church, working at around the same time as Turing, described the theory of *recursive* functions. Similarly, Emile Post, also a contemporary of Turing and Church, and, like Church, an American, described in 1936 a related class of abstract machines. These consisted of an infinite tape on which an abstract problem-solving mechanism could either place a mark or erase a mark. The result of a computation was the marked portion of the tape on completion of the process. These formalisms, and subsequent developments such as Wang's machines (like TMs but with an alphabet of only blank and "*"), and the register machines of Shepherdson and Sturgis (closely resembling a simple assembler code-based system but with registers that can store arbitrarily large integers), have been shown to be:

a) equivalent in computational power to Turing machines, and therefore,

b) more powerful than any machine with limited storage resources.

Turing's thesis concerns the computational power of the primitive machines which he defined, which we call *Turing machines*. Essentially, in its broadest sense, Turing's thesis says this:

---

[1] Reading related to the work referenced in this section can be found in the "Further Reading" section.

> *Any well-defined information processing task can be carried out by some Turing machine.*

However, in a mathematical sense, the thesis says:

> *Any computation that can be realised as an effective procedure can also be realised by some Turing machine.*

This last version of the thesis is probably not particularly helpful, as we have yet to specify what constitutes an *effective procedure*. Turing was working in an area of pure mathematics which had, for many years studied the processes of "doing" mathematics itself, i.e. the procedures that were used by mathematicians to actually carry out the calculations and reasoning that mathematicians do. At the time Turing formulated the notion of TMs, the "computers" that his machines were intended to emulate were thus *humans engaged in mathematical activity*. An effective procedure, then, is a step-by-step series of unambiguous instructions that carry out some task, *but specified to such a level of detail that a machine could execute them.* In this context, as far as we are concerned, Turing's thesis tells us this:

> *There is no machine (either real or abstract) that is more powerful, in terms of what it can compute, than the Turing machine.*

Therefore:

> *The computational limitations of the Turing machine are the least limitations of any real computer.*

Turing machines are radically different from the computers with which you and I interact in our everyday lives. These real computers have *interfaces*; they may display detailed graphics, or even animations and video, or play us music, or enable us to speak to, or play games with, them, or with other people using them. A small computer may run an artificial intelligence program, such as an expert system that solves problems that even twenty years ago would have been impossible on anything but the largest mainframe. Yet, in a formal sense, everything that happens on the *surface*, so to speak, of the computer system, is merely a *representation*, an alternative encoding, of the results of symbol manipulation procedures. Moreover, these procedures can be modelled by a simple machine that merely repeatedly reads a single symbol, writes a single symbol, and then moves to examine the next symbol to the right or left. What's more, this simple abstract machine was formulated at a time when "real computers" were things like mechanical adders and Hollerith card readers.

The Turing machine represents a formalism that allows us to study the nature of computation in a way that is independent of any single manufacturer's architecture, machine languages, high level programming languages or storage characteristics and capacity. TMs are thus a single formalism that can be applied to the study of topics such as the complexity of automated processes and languages (which is how we view them in the remainder of this part of the book).

In terms of languages, we have already seen the close relationship between TMs and the Chomsky hierarchy, and we refine this later in this chapter, and continue to do so in the next chapter. In Chapter 12, we consider in detail the notion of complexity as it applies both to TMs and to programs on real machines. However, an even more exciting prospect is suggested by our statement that the limitations of the TM are the least limitations of any real computer. For a start, only a real computer to which we could indefinitely add additional storage capacity as required could approach the power of the TM. It is these limitations, and their ramifications in the real world of computing, that are the main concern of this part of the book.

In order to support the arguments in this chapter, we define a single Turing machine that can model the behaviour of any other TM. We call this machine the *universal* TM (*UTM*). The existence of *UTM* not only provides further evidence in support of Turing's thesis, that the Turing machine is the most powerful computational machine, but also plays a part in finally establishing the relationship between computable languages and abstract machines. In the course of this, *UTM* also features in the next chapter, in a demonstration of an extremely significant result in computer science: the general unsolvability of what is called the *halting problem.*

Our *UTM* will accept as its input:

1. (a coded version of) a TM, M

2. (a coded version of) the marked portion of an input tape to M.

UTM then simulates the activities that M would have carried out on the given input tape.

Since *UTM* requires a coded version of the machine, M, and the input to M, our next task is to define a suitable scheme for deriving such codes.

10.3 Coding a Turing Machine and its Tape as a Binary Number

We first consider a simple method of representing any TM as a binary code. We then see how the code can be used to code any input tape for the coded machine. The coded machine and the coded input can then be used as input to *UTM*.

10.3.1 Coding Any TM

To describe the scheme for coding any TM as a string of 1 s and 0 s, we will use an example TM. As we proceed, you should appreciate that the coding scheme could be applied to any TM, consisting of any number of states and using any size of alphabet.

Our example machine, M, is shown in Figure 10.1.

An alternative representation of M (Figure 10.1) is shown in Table 10.1. You will note that each instruction (initial state, read symbol, write symbol, direction, and end state) is represented as a five-element list. In mathematical parlance, a five-element list is known as a *quintuple*. Table 10.1 thus provides the quintuples of the TM, M. This representation of TMs is fairly common. We use it here as it supports the subsequent discussion in this chapter and the following two chapters.

We begin by associating the *states* of M (Figure 10.1) with strings of 0s as shown in Table 10.2. It should be clear that we could code the states of any TM using the scheme.

Next, as shown in Table 10.3, we code the *tape symbols (alphabet)* of the machine, which, for our machine M, is $\{a, b, C, X, Y, Z\}$ and the *blank* symbol. Again, we can see that this scheme could apply to a TM alphabet of any number of symbols.

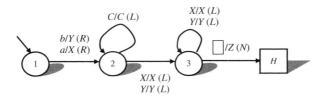

Figure 10.1 A simple Turing machine.

Table 10.1 The *quintuple* representation of the Turing machine in Figure 10.1.

$(1, a, X, R, 2)$
$(1, b, Y, R, 2)$
$(2, C, C, L, 2)$
$(2, X, X, L, 3)$
$(2, Y, Y, L, 3)$
$(3, X, X, L, 3)$
$(3, Y, Y, L, 3)$
$(3, \square, Z, N, H)$

Table 10.2 Coding the states of the machine of Figure 10.1 as strings of zeros.

State	Code	Comment
1	0	*Start* state of any machine always coded as 0
H	00	*Halt* state of any machine always coded as 00
2	000	For the remaining states, we simply assign to each a unique string of 0 s greater than 2 in length
3	0000	

Table 10.3 Coding the alphabet of the machine of Figure 10.1 as strings of zeros.

Symbol	Code	Comment
□	0	*Blank* always coded as 0
a	00	For the rest of the alphabet, we simply assign to each any unique string of 0s greater than 1 in length
b	000	
C	0000	
X	00000	
Y	000000	
Z	0000000	

We also code the direction symbols L, R and N, as shown in Table 10.4.

You will note that some of the codes representing the states (Table 10.2) are identical to those used for the alphabet (Table 10.3) and direction symbols (Table 10.4). However, this does not matter; as we will see in a moment, our coding scheme ensures that they do not become mixed up.

We have coded the states, the alphabet, and the direction symbols. We now code the instructions (*quintuples*) themselves. Figure 10.2 shows how to code a particular quintuple of M. Once again, it is clear that any quintuple could be coded in this way.

Now, we simply code the whole machine as a sequence of quintuple codes, each followed by 1, and we place a 1 at the start of the whole code. The scheme is outlined in Figure 10.3.

The above should be sufficient to convince you that we could apply such a coding scheme to any TM, since any TM must have a finite number of states, tape symbols and quintuples.

Table 10.4 Coding a Turing machine's direction symbols as strings of zeros.

Symbol	Code
L	0
R	00
N	000

Figure 10.2 How one *quintuple* of the TM from Figure 10.1 is represented as a binary code (for coding scheme see Tables 10.2 to 10.4).

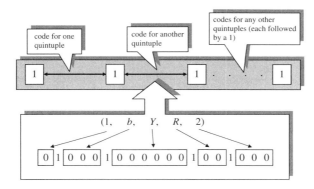

Figure 10.3 How a whole Turing machine is coded as a binary number (the example *quintuple* is that of Figure 10.2).

Such a binary code for a Turing machine could be presented as the input to any TM that expects arbitrary binary numbers as its input. For example, we could present the code for the above machine, *M*, to the *MULT* TM (described in Chapter 9) as one of the two numbers to be multiplied. We could even present the coded version of *MULT* to the *MULT* machine as one of its input numbers, which would mean that *MULT* is being asked to multiply its own code by another number! All this may seem rather strange, but it does emphasise the point that codes can be viewed at many different levels. The code for *M* can be regarded as another representation of *M* itself (which is how *UTM* will see it), as a *string* in $\{0, 1\}^+$ (a non-empty string of 0 s and 1 s) or as a *binary number*. This is essentially no different from seeing Pascal source code as text (the text editor's view) or as a program (the compiler's view).

Since our *UTM* will require not only the coded version of a machine, but also a coded version of an input to that machine, we need to specify a scheme for coding any marked tape of the machine we are coding.

Table 10.5 Coding a Turing machine's input tape as a binary number.

Code	Application to an input tape for the TM of Figure 10.1
We use the same codes for the tape symbols as we use for the machine's alphabet (Table 10.3). As for the quintuples (Figure 10.2), we place a 1 after the code for each symbol. We place a 1 at the beginning of the coded input.	

10.3.2 Coding the Tape

The coding scheme for the input is described in Table 10.5. Note that for a given machine, we use the same codes for the symbols on the input tape as those used to code the machine's alphabet (Table 10.3).

It should be clear that any input tape to any coded machine could be similarly represented using the scheme in Table 10.5.

We have now seen how to take any TM, M, and code it, and any of its input tapes, with a coding scheme that requires only the alphabet $\{0, 1\}$. In the next section, we use our coding scheme as an integral component of the architecture of our universal Turing machine, UTM.

10.4 The Universal Turing Machine

UTM is described here as a 3-*tape* deterministic TM. Though our TMs have hitherto possessed only one tape, it is quite reasonable for a TM to have any (fixed) number of tapes. We discuss such machines in more detail a little later in this chapter, when we also show that a "multi-tape" TM can be modelled by an equivalent single-tape machine. For the moment, it suffices to say that in a multi-tape TM, each tape is numbered, and each instruction is accompanied by an integer which specifies the number of the tape on which the read, write and head movement operations in that instruction are to take place.

The 3-tape UTM receives as its input a coded version of a TM and essentially simulates the behaviour of the coded machine on a coded version of input to that machine. UTM therefore models the behaviour of the original machine on the original input, producing as its output a coded version of the output that would be produced by the original machine, which can then be decoded.

You may at this point be concerned that there is some "trickery" going on here. You might say that everything that *UTM* does is in terms of manipulating *codes*, not the "real" thing. If you *do* think that it's trickery, then it can be pointed out that every single program you have ever written is part of the same type of "trickery", as

- the source file,

- the input,

and

- the output

are all similarly coded and subsequently decoded when your program is run. So if you are not prepared to believe in the representational power of coding schemes, you must believe that no program you have ever used does anything useful!

Next, we specify the initial set up of *UTM*'s three tapes. Following this, we briefly consider the operation of *UTM*. You will appreciate that the construction of *UTM* described would be capable of simulating the operation of any TM, *M*, given any suitable input tape to *M*.

10.4.1   *UTM*'s Tapes

Our version of *UTM* is configured so that tape 1 is as shown in Figure 10.4.

As can be seen from Figure 10.4, tape 1 contains the code for the machine *M*, followed immediately by the code for *M*'s input.

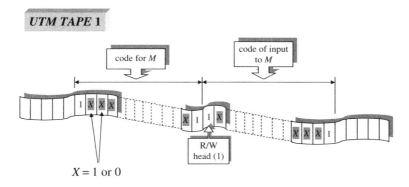

Figure 10.4 The set up of tape 1 of *UTM*. The machine, *M*, is coded as specified in Figure 10.3. The input to *M* is coded as specified in Table 10.5.

Figure 10.5 The set up of tape 2 of *UTM*. *UTM* initially copies the coded input (see Figure 10.4) from tape 1 onto tape 2. Tape 2 is then used as a "work tape" by *UTM*.

UTM will use its second tape as a "work tape", on which to simulate *M*'s operation on its input by carrying out the operations specified by the coded quintuples on the coded data. Thus, the first thing *UTM* does is to copy the coded input from tape 1 to tape 2, at which point, tape 2 will be as specified in Figure 10.5.

Tape 3 of *UTM* will be used only to store a sequence of 0 s denoting, at any stage, the current state of *M*. The organisation of tape 3 is depicted in Figure 10.6.

To summarise:

tape 1 stores the coded machine *M*, followed by the coded input to *M*
tape 2 is used by *UTM* to work on a copy of *M*'s coded input
tape 3 is used to hold a sequence of 0 s representing *M*'s current state.

10.4.2   The Operation of *UTM*

UTM operates as specified by the procedure in Table 10.6. I leave it to you to appreciate how the procedure itself could be represented as a TM. For clarification of the procedure in Table 10.6, you are advised to refer to the coding scheme and the configurations of *UTM*'s three tapes described above.

A *UTM* based on the procedure in Table 10.6 is capable of carrying out the computation specified by any TM, given an initial set up as described above. As

Figure 10.6 The set up of tape 3 of *UTM*. Tape 3 is used to hold the code (a sequence of zeros) for *M*'s current state. It is initialised to a single 0, as any machine's start state is coded as 0.

input to such a machine, we code the TM according to our scheme, and then we can code an input tape to that TM using the same scheme. *UTM* will then simulate the computation performed by the original machine.

In particular:

- If *M*, the original machine, would produce a solution given a particular input tape, then *UTM* would produce a coded version of that solution (on its tape 2).

- If *M* would reach a state, *s*, in which it had no applicable quintuples for the current state and tape symbols, then *UTM* would halt with the code for the state *s* on tape 3, and tape 2 would contain a coded version of *M*'s tape configuration at the time *M* stopped.[2]

- If *M* would reach a situation where it looped indefinitely, repeating the same actions over and over again, then *UTM* would also do the same. This is very significant and highly relevant to a discussion in the next chapter.

We can now appreciate that we have described a 3-tape deterministic TM, *UTM*, that can, if we accept Turing's thesis, carry out any *effective procedure*. We will see later that from our 3-tape version we could construct a deterministic "standard" single-tape *UTM*.

---

[2] This assumes that the "while" loop in Table 10.6 terminates if no suitable quintuple can be found.

Table 10.6 The operation of the universal Turing machine, *UTM*, in simulating the behaviour of a TM, *M*. For details of the coding scheme used, see Tables 10.2 to 10.5, and Figures 10.2 and 10.3. For the set up of *UTM*'s three tapes, see Figures 10.4 to 10.6.

Operation of UTM	Comments
1 Copy the code for *M*'s data from tape 1 to tape 2	Tape 2 is to be used as a work tape by *UTM*
2 Write a 0 on tape 3 to represent *M*'s start state	0 represents *M*'s *start* state
3 while tape 3 does not contain exactly 2 zeros do	00 represents *M*'s *halt* state
find a quintuple of *M* on tape 1 for *current state* on tape 3	*UTM* proceeds rightwards along tape 1, comparing the sequence of zeros in each quintuple of *M* that represent a current state, with the sequence of zeros on tape 3
if quintuple found then	
if *current symbol* code on tape 2 = quintuple *read symbol* code on tape 1	*UTM* needs to compare sequences of zeros to do this
then	
on tape 2, overwrite the *current symbol* code with the quintuple *write symbol* code on tape 1	*UTM* may have to *shuffle* symbols, since each alphabet symbol is represented by a different length string of zeros (Chapter 7 contains examples of this type of activity)
move the read head on tape 2 according to the quintuple *direction* code on tape 1	*UTM* must move its tape 2 read/write head to the next 1 to the left or right (or not at all)
overwrite the *state* code on tape 3 with the quintuple *next state* code from tape 1	*UTM* replaces the string of zeros on tape 3 with the string of zeros representing the next state of *M*
endif	
endif	
endwhile	

10.4.3   Some Implications of *UTM*

The existence of *UTM* makes Turing's thesis that TMs are the pinnacle of functional power even more compelling. We have seen that a TM that can "run", as a kind of "program", any other TM needs to be no more sophisticated than a standard TM. In a sense, *UTM* actually embodies the *effective procedure* of *carrying out the effective procedure* specified by another Turing machine.

Other designs of *UTM* may be suggested, based on different schemes for coding the machine *M* and its input. All are equivalent. *UTM* is a *Turing machine* and can itself be coded according to the scheme for one of the other *UTM*s. The code can then be presented, along with a coded version of an input tape, as input to that other *UTM*. In that sense also, *UTM* is truly *universal.*

In a more practical vein, *UTM* is a closer analogy to the stored program computer than the individual TMs we have considered so far. Our other TMs are essentially purpose-built machines that compute a given function. *UTM*, with its ability to carry out computations described to it in some coded form, is a counterpart of real computers that carry out computations described to them in the form of codes called *programs*. However, once again it must be emphasised that a given computer could only be *as* powerful as *UTM* if it had unlimited storage capabilities. Even then, it could not be *more* powerful.

In the next section, we provide even more compelling evidence in support of Turing's thesis. In doing so, we close two issues that were left open earlier in this book. Firstly, in Chapter 7 we saw how a non-deterministic TM is the recogniser for the type 0 languages. Secondly, earlier in this chapter, we encountered a 3-tape *UTM*. We now see that the additional power that seems to be provided by such features is illusory: the basic, single-tape TM is the most powerful computational device.

10.5 Non-deterministic TMs

In Chapter 7, we saw that the *non-deterministic* TM is the abstract machine counterpart of the *type* 0, or *unrestricted* grammars, the most general classification of grammars of the Chomsky hierarchy. We saw a method for constructing non-deterministic TMs from the productions of type 0 grammars. Now we return to the notion of non-deterministic TMs to demonstrate that non-determinism in TMs provides us with no extra *functional* power (although we see in Chapter 12 that things change if we also consider the *time* that it might take to carry out certain computations). This result has implications for *parallel computation*, to which we return later in this chapter. Here, only a sketch is given of the method for converting any non-deterministic TM into an equivalent deterministic TM. However, some of the techniques introduced are fundamental to discussions in the remainder of this part of the book.

Now, if a TM is non-deterministic and it correctly computes some function (such as accepting the strings of a language), that TM will be capable of reaching its halt state, leaving some kind of result on the tape. For a non-deterministic machine this is alternatively expressed as "there is some sequence of *transitions* that would result in the machine reaching its halt state from one of its start states". In terms of our discussion from earlier in this chapter, the phrase

"there is some sequence of *transitions*"

can be alternatively expressed as

"there is some sequence of *quintuples*".

Suppose, then, that we have a non-deterministic TM that sets off in attempting to solve some problem, or produce some result. If there *is* a result, and it can try all possible applicable sequences of quintuples *systematically* (we will return to this in a moment), then it will find the result. What we are going to discover is that there is a *deterministic* method for trying out every possible sequence of quintuples (shorter sequences first) and stopping when (or rather *if*) a sequence of quintuples is found that leads to a halt state of the original machine. Moreover, it will be argued that a *deterministic* TM could apply the method described. Analogously to *UTM*, the construction described represents a case where one TM executes the instructions of another, as if the other TM was a "program".

In order to simplify the description of our deterministic version of a non-deterministic TM, we shall assume that we have a special TM with four tapes. We will then show how multi-tape machines can be modelled by single-tape machines, as promised earlier.

10.6 Converting a Non-deterministic TM into a 4-tape Deterministic TM

We now describe the conversion of a non-deterministic TM into a 4-tape deterministic TM. Let us call the non-deterministic TM, N, and the corresponding deterministic 4-tape machine, D.

10.6.1 The Four Tapes of the Deterministic Machine, *D*

We begin by describing the organisation of each of *D*'s four tapes, beginning with tape 1, which is shown in Figure 10.7.

As shown by Figure 10.7, tape 1 of *D* stores:

- a copy of the marked portion of the tape that *N* would have started with (we can assume for the sake of simplicity that *N*'s computation would have begun with its read/write head on the leftmost symbol of its input)

- the label of *N*'s current state, initially marked with *N*'s start state label (let us assume that the start state is always labelled *S*)

- the label of *N*'s halt state (we can assume that *N* has only one halt state; cf. exercise 1, Chapter 7), which in Figure 10.7 is *H*.

Note that we assume that labels such as state names and alphabet symbols each occupy one *square* of a tape, i.e., are *single symbols*. In the formal world, there is no concept of a "character set" in which we "run out" of single-symbol labels; there is a potentially infinite number of available single-symbol labels. Thus, a TM of any size can be assumed to require state and alphabet labels each consisting of one symbol.

Now we can describe tape 2, schematically represented in Figure 10.8.

Figure 10.8 tells us that tape 2 of *D*, the deterministic 4-tape TM, contains the quintuples of *N*, each occupying five consecutive tape squares, each preceded

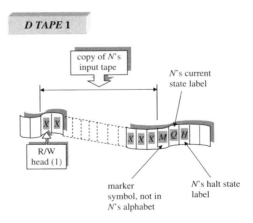

Figure 10.7 The set up of tape 1 of the deterministic, 4-tape TM, *D* (a deterministic version of a non-deterministic TM, *N*).

D TAPE 2

N's quintuples, each preceded
by a unique symbol not in N's
alphabet

a X X X X X b X X X X X

R/W
head (2)

one quintuple of N

Figure 10.8 The set up of tape 2 of the deterministic, 4-tape TM, D (a deterministic version of a non-deterministic TM, N). Tape 2 stores the quintuples of N.

by a unique single symbol that is not a symbol of N's alphabet. To illustrate, Figure 10.9 shows how a quintuple would be coded on tape 2 of D.

Next, tape 3 of D. Tape 3 is where D will repeatedly perform the systematic application of N's quintuples, starting from a copy of N's input tape (from tape 1, specified in Figure 10.7). D's first activity, then, is to set up *tape* 3 as shown in Figure 10.10.

Note that D places special "markers" at either end of the copied input. These are necessary, as after the application of a sequence of N's quintuples that leads to a dead end, D will have to erase the marked portion of tape 3, and copy the original input from tape 1 again. The potential problem is that N's computation may result in blank squares in between marked symbols, resulting in a tape that is no longer *packed*, so D needs a way of detecting the extreme ends of the marked portion of the tape.

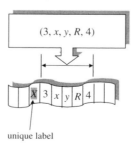

$(3, x, y, R, 4)$

X 3 x y R 4

unique label

Figure 10.9 An example quintuple coded on tape 2 of N (see Figure 10.8).

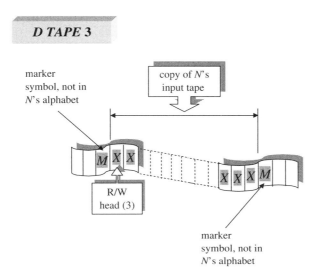

Figure 10.10 The set up of tape 3 of the deterministic, 4-tape TM, D. On tape 3, D will systematically apply the quintuples of the non-deterministic machine, N.

The existence of the markers causes a slight complication in the operation of D. One of N's quintuples may result in D's tape 3 head moving onto the marker. When this happens, D must simply move the marker one square to the right or left, depending on whether the marker was encountered on a right or left move, respectively. D must replace the marker by a blank, and then overwrite the blank on the right or left, respectively, with the marker. D must also then move the head back one square to where the marker originally was (this corresponds to the position in the tape expected by the next applicable quintuple of N). Every time D makes a left or right move on tape 3, then it must check if the next symbol is the marker; if so, it must take the appropriate action before attempting to apply the next quintuple of N.

10.6.2 The Systematic Generation of the Strings of Quintuple Labels

The whole process hinges on the statement, made more than once above, that D will apply the quintuples of N *in a systematic way*. As N can consist of only a *finite* number of quintuples, the set of symbols used to label the quintuples of N (on tape 2), is an *alphabet*, and a sequence of symbols representing labels for some sequence of quintuples of N is a string taken from that alphabet. So, the set of all possible sequences of quintuples that we may have to try, starting from the initial tape given to N, is denoted by the set of all possible non-empty strings formed from symbols of the alphabet of labels.

Let us call the alphabet of quintuple labels Q. Recall from Chapter 2 that the set of every possible non-empty string formed using symbols of the alphabet Q is denoted by Q^+. Now, Q^+, for any alphabet Q, is not only *infinite* but also *enumerable*. Q^+ is a *countably infinite set*. Now, by *Turing's thesis*, if a *program* can enumerate Q^+, then a TM can do it.[3] That is exactly what we need as part of D:

- A sub-machine that will generate sequences of quintuple labels, one by one, shortest sequences first, so that the sequence of N's quintuples that is denoted by the sequence of symbols can be executed by D. Moreover, we must be sure that every possible sequence is going to be tried, if it proves necessary to do so.

To be prepared to try every possible sequence of quintuples is a rather brute-force approach, of course. Many (in fact probably most) of the strings of labels generated will not denote "legal" sequences of N's quintuples for the given tape, or for *any* tape at all. Moreover, the method will try any sequence that has a prefix that has already led nowhere (there are likely to be an infinite number of these, also). However, because D tries the shorter sequences before the longer ones, at least we know that if there *is* a solution that N could have found, our machine D will eventually come across the sequence of quintuples that represents such a solution. We should also point out that if there is not a solution, then D will go on forever trying different sequences of quintuples (but then, if a solution does not exist, N would also be unable to find it!).

So, we finally come to tape 4 of D. Tape 4 is where D is going to repeatedly generate the next sequence of quintuple labels (strings in the set Q^+) that are going to be tried. Only an overview of one method will be given here. An exercise at the end of the chapter invites you to construct a TM to achieve this task.

To demonstrate the method, we will assume an alphabet $\{a, b, c\}$. The approach will be:

generate all strings of length 1:

$$a, b, c$$

generate all strings of length 2:

$$aa, ab, ac, ba, bb, bc, ca, cb, cc$$

---

[3] A pseudo-code program was given in Chapter 2 to perform the same task.

generate all strings of length 3:

$$aaa, aab, aac, aba, abb, abc, aca, acb, acc, baa, bab, bac, bba, bbb,$$
$$bbc, bca, bcb, bcc, caa, cab, cac, cba, cbb, cbc, cca, ccb, ccc$$

You may notice from the order in which the strings are written that we obtain all of the strings of length n by copying each of the strings of length $n - 1$ three times (as there are three alphabet symbols). We then append a different symbol of the alphabet to each one. For example, from aa we generate aaa, aab and aac.

Let us assume that the machine initially has the three alphabet symbols somewhere on its tape. To generate the three strings of length 1 it simply copies each symbol to the right, each being followed by a *blank*, with the final blank followed by some marker (say "*").

The blanks following each of the strings in Figure 10.11 can be used both to separate the strings and to enable an auxiliary marker to be placed there, thus allowing the machine to keep track of the current string in any of its operations.

Figure 10.11 Tape 4 of D, when the set of quintuple labels is $\{a,b,c\}$, and all the strings of length equal to 1 have been generated. This is part of D's systematic generation of all strings in $\{a, b, c\}^{+}$.

The next stage, shown in Figure 10.12, involves copying the first string of length 1 (*a*) three times to the right of the *. The number of times the string is copied can be determined by "ticking off", one by one, the symbols of the alphabet, which, as shown in Figure 10.11, is stored somewhere to the left of the portion of tape being used to generate the strings.

From the situation in Figure 10.12, the next stage is to append each symbol of the alphabet, in turn, to the three copied *a*s, giving the situation in Figure 10.13.

The next string of length 1, (*b*) is then copied three times to the right, as illustrated in Figure 10.14.

From Figure 10.14, each symbol of the alphabet is appended, in turn, to the strings just copied, yielding the situation in Figure 10.15.

The next stage from Figure 10.15, is to copy the next string of length 1 (*c*) three times to the right, leading to the set up shown in Figure 10.16.

From Figure 10.16, each symbol of the alphabet is appended to each string that was copied in the previous stage, giving a situation shown in Figure 10.17.

Now, when the machine returns along the tape, then attempts to copy the next string of length 1, it encounters the "*". The machine thus detects that it has

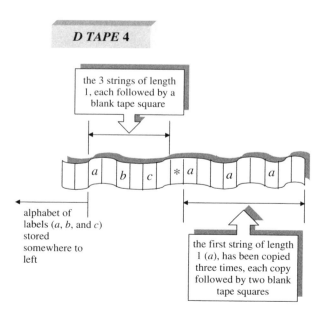

Figure 10.12 On tape 4, *D* is generating all strings in $\{a, b, c\}^+$ of length 2. The first stage is to copy the first string of length 1 (Figure 10.11) three times (since the alphabet $\{a, b, c\}$ contains three symbols).

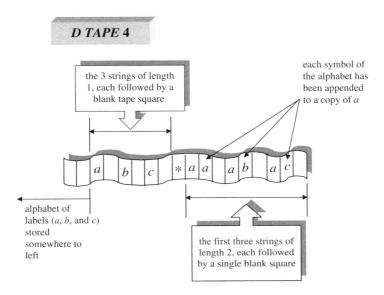

Figure 10.13 Tape 4 of D, when D is generating all strings in $\{a, b, c\}^+$ of length 2. D has now appended a symbol of the alphabet to each copy of a from Figure 10.12.

finished generating all strings of length 2. The machine can place a "*" at the end of the strings it created (Figure 10.17), as depicted in Figure 10.18.

From Figure 10.18, the machine is ready to begin the whole process again, i.e. copying each string between the two rightmost *s three times to the right, each time appending each symbol of the alphabet in turn to the newly copied strings, and so on.

Of course, it will be necessary for D to try the quintuple sequences each time a certain number of them have been generated (since the generation process can go on forever). It could, for example, try each new sequence as it was created, i.e., each time a symbol of the alphabet is appended to a recently copied sequence. Alternatively, it could generate all of the sequences of a given length, and then try each of those sequences in turn.

10.6.3 The Operation of D

Here then, is a sketch of the method used by the 4-tape TM, D, to deterministically model the behaviour of the non-deterministic TM, N.

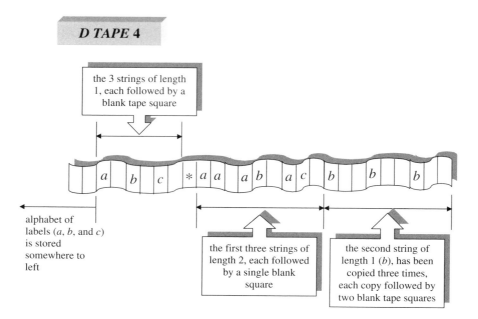

Figure 10.14 Tape 4 of D, when it is generating all strings in $\{a, b, c\}^+$ of length 2. From the situation in Figure 10.13, D has now added 3 copies of the second string of length 1 (b).

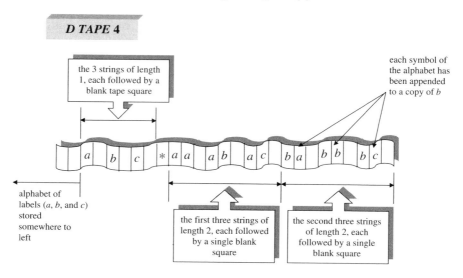

Figure 10.15 Tape 4 of D, when it is generating all strings in $\{a, b, c\}^+$ of length 2. D has now appended a symbol of the alphabet to each copy of b from Figure 10.14.

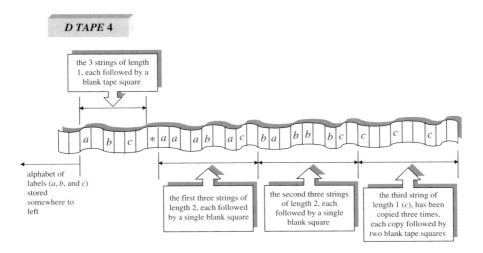

Figure 10.16 Tape 4 of D, when it is generating all strings in $\{a,\,b,\,c\}^+$ of length 2. From the situation in Figure 10.15, D has now added 3 copies of the second string of length 1 (c).

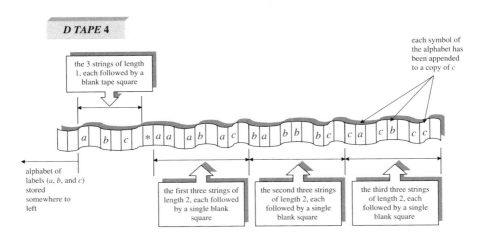

Figure 10.17 Tape 4 of D, when it is generating all strings in $\{a,\,b,\,c\}^+$ of length 2. D has now appended a symbol of the alphabet to each copy of c from Figure 10.16.

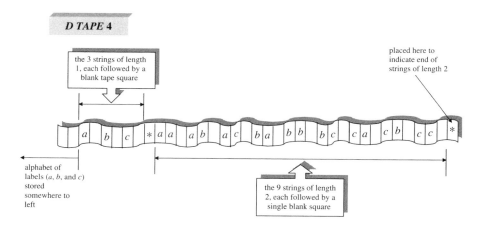

Figure 10.18 D has now generated all strings in $\{a,\,b,\,c\}^{+}$ of length 2, and has now appended a "*" to the end of the portion of tape containing the strings.

D is assumed to be initially configured as described earlier, i.e.:

Tape 1 (Figure 10.7) contains a copy of N's input, and its current halt state label,

Tape 2 (Figure 10.8) contains the labelled quintuples of N,

Tape 3 (Figure 10.10) is initially blank, and

Tape 4 contains the quintuple label symbols.

D operates as shown in Figure 10.19.

The most complicated process in the method specified in Figure 10.19 is the box that begins "Try sequence of quintuples . . .". This is expanded in Figure 10.20.

Sufficient detail has been presented for you to appreciate that the above construction could be applied to any similar situation.

10.6.4   The Equivalence of Non-deterministic TMs and 4-Tape Deterministic TMs

Thus far, you are hopefully convinced that any non-deterministic TM, N, can be represented by a 4-tape deterministic TM, D. D and N are equivalent, in the sense that, given any input tape to N, and assuming that D's tapes are initially configured as described above, if N could have found a solution then so will D, and if N would never find a solution then neither will D.

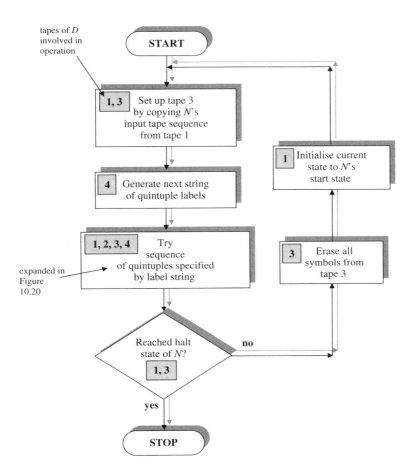

tapes of D
involved in
operation

expanded in
Figure
10.20

Figure 10.19 The operation of the 4-tape TM, D, that deterministically models the behaviour of a non-deterministic machine, N.

We now turn our attention to establishing that a multi-tape deterministic TM can be modelled by an equivalent single-tape machine.

10.7 Converting a Multi-tape TM into a Single-tape TM

A *multi-tape TM*, as described above, is a TM with two or more tapes, each of which has its own read/write head. The instructions on the arcs of such TMs contain an extra designator, this being the number of the tape on which the next

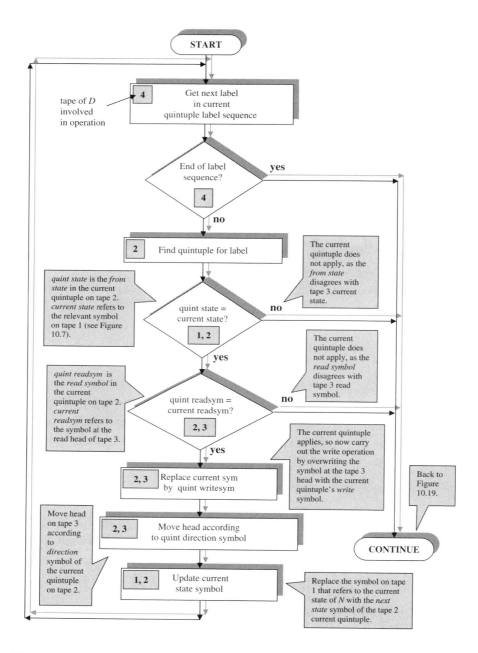

Figure 10.20 Expansion of the box that begins "Try sequence of quintuples . . ." from Figure 10.19. How D models the behaviour of N specified by its quintuples.

operation is to take place. Thus, in place of quintuples we have *sextuples*. Any such machine can be modelled by a single-tape TM, as will now be demonstrated.

Let us call the multi-tape machine M and the resulting single-tape machine S. The conversion method we consider here consists of representing the symbols from M's many input tapes on S's single tape in such a way that a sextuple of M can be represented as several *quintuples* of S. We will use an example involving the conversion of a 3-tape machine into a single-tape machine to demonstrate the method.

10.7.1   Example: Representing Three Tapes as One

Let us suppose we have a 3-tape machine, M, with its tapes initially set up as in Figure 10.21.

We use a scheme that systematically places the symbols from the three tapes of M on the single tape of S. From the three tapes above, we present the tape to S that is shown in Figure 10.22.

Thus, from left to right, and taking each tape of M in turn, we place the leftmost symbol from tape 1, followed by a blank (or a special marker if this symbol was pointed at by the read/write head of tape 1). This is followed by the leftmost symbol from tape 2, followed by a blank (or a special marker, if necessary,

Figure 10.21 Example configurations of the three tapes of a hypothetical 3-tape Turing machine.

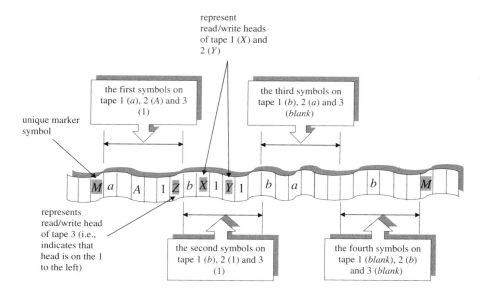

Figure 10.22 Representing the three tapes of Figure 10.21 on a single tape.

representing the read/write head of tape 2), and so on. It will be noted that tape 3 had the least marked squares, so pairs of blanks are left where tape 3 symbols would be expected on S's tape. Similarly, for tape 1, two blanks were left so that the final symbol of tape 2, the tape of M with the most marked squares, could be placed on the single tape.

The end of input tape markers, right and left, and the three read/write head markers (X, Y and Z, for tapes 1–3, respectively) are required to be symbols that are not in the alphabet of M.

The encoding of the three tapes of M as a single tape of S represented in Figure 10.22 captures all of the information from the original three tapes. For example, consider the seventh tape square to the right of the left-hand marker symbol. This represents the second occupied tape square of tape 1. It contains a b, which tells us the second tape square from the left of tape 1 of M should also contain a b (as it does). Moreover, the square to the right of the b in question, on S's tape, contains X, the read/write head marker representing the head position on tape 1 of M. This tells us that the read/write head of M's tape 1 should be pointing at the second symbol from the left, i.e., the first b from the left (as it is). All other information from the initial three tapes can be similarly recovered from our single-tape representation.

10.7.2 The Operation of the Single-tape Machine, S

In the machine S, *single* instructions (sextuples) of the multi-tape machine, M, become *several* instructions (quintuples) of the single-tape machine. These quintuples operate on a single-tape set up representing the multiple tapes of M, as specified in Figure 10.22.

Consider as an example the sextuple $(1, a, A, R, 2, 2)$. The meaning of this sextuple is:

> "when in state 1, if the symbol at the read/write head of the current tape is a, replace it by A, move right one square on the current tape, switch processing to tape 2 and move into state 2".

Let us assume that the alphabet of tape symbols of S is $\{a, b, A, 1\}$, and that (as in Figure 10.22) the markers used on the single tape to represent the read/write heads on the three tapes, tapes 1, 2 and 3, are X, Y and Z, respectively. For the corresponding single-tape machine S, assuming that its read/write head is currently located at the tape square representing the appropriate marked square of M's tape 1, the sextuple $(1, a, A, R, 2, 2)$ is interpreted by the single-tape machine as explained in Table 10.7.

Every single instruction (i.e. *sextuple*) of the multi-tape machine has to be represented in a similar way to that described above. For a *left* move instruction, a very similar construction could be used, except that the single-tape machine would have to move *left* six squares from the symbol to the left of the current tape's read/write head marker. For a *no* move instruction, movement of the read/write head marker would be unnecessary. For a sextuple that specified that the

Table 10.7 An overview of how a single-tape TM, S, simulates the execution of the sextuple $(1, a, A, R, 2, 2)$ of a 3-tape TM, M. The three tapes of M are assumed to have been coded onto a single tape, as described in Figures 10.21 and 10.22. The *current state* of M is assumed to be 1.

M instruction $(1, a, A, R, 2, 2)$	Corresponding behaviour of S
"If the symbol at the read/write head of tape 1 is a, replace it by A ...	Since (see text), the read/write head of S is pointing at the symbol on its tape that represents the current symbol on M's tape 1, if that symbol is a, S replaces it with A.
... move right one square on tape 1, ...	We have three tapes of M coded onto one tape of S. Each tape square of M is represented by two tape squares of S (one for the symbol, one to accommodate the marker that indicates the position of one of M's read/write heads). Thus, moving "one square" on a tape of M involves moving *six* squares on S's tape. S also has to move the marker (X) representing M's tape 1 read/write head so that it is in the appropriate place (i.e., next to the symbol representing one move right on M's tape 1).
... switch processing to tape 2 and move into state 2."	S must search for the marker that represents the read/write head of M's tape 2 (in our example, Y).

next operation was to take place on the same tape, there would be no need to search for the marker representing the read/write head of another tape.

Like our 4-tape deterministic machine D earlier in the chapter, S uses "marker" symbols to indicate the extremes of the marked portion of its tape. Thus, like D, S has to check for this "end of tape" marker each time it moves its head to the right or left in the portion of tape representing the tapes of the multi-tape machine (see Figure 10.22). Such situations correspond to those wherein the multi-tape machine would move its head onto a blank square outside of the marked portion of one of its tapes. In such circumstances, S must *shuffle* the "end of tape" marker to the left or right by $2k$ squares, where k is the number of tapes of the simulated machine, M. For example, a "move right" on a single-tape TM representing a 10-tape TM would involve moving 20 squares to the right (shuffling the "end of tape" marker right appropriately if necessary), and we would of course need 10 distinct read/write head markers, and so on.

You should be able to see how the above could be adapted for multi-tape machines with any number of tapes. You are asked to add more detail to the construction in an exercise at the end of this chapter.

10.7.3 The Equivalence of Deterministic Multi-tape TMs and Deterministic Single-tape TMs

The preceding discussion justifies the following statement:

> *Any deterministic multi-tape TM,M, can be replaced by a deterministic single-tape TM, S, that is equivalent to M.*

In the preceding discussion, we have seen that it can be a convenience to define multi-tape, rather than single-tape, machines. Recall also the previous chapter, when we described TMs such as $MULT$ and DIV. These machines featured as "sub-machines" other TMs we had defined, such as ADD and $SUBTRACT$. We were forced to use different areas of the tape to deal with the computations of different sub-machines, which led to some tricky copying and shuffling. A multi-tape approach would have enabled us to use certain tapes in a similar way to the "registers" or "accumulators" of the CPU, to store temporarily the results of sub-machines that would be subsequently needed.

A second, and related, feature of multi-tape machines is that they permit a *modularity* of "memory organisation", a memory "architecture", so to speak, which has more in common with the architecture of the real computer system that the single-tape TM. This can be appreciated by considering TMs such as the deterministic multi-tape machines we specified in the previous section, where the *program* (i.e. the quintuples of another machine) was stored on one tape and the *data* were stored on another. Moreover, the machine used a further tape as a "working memory" area, and another tape as a kind of *program counter* (the tape where the sequences of quintuple labels were generated).

The modularity facilitated by a multi-tape TM architecture can be useful for specifying complex processes, enabling what is sometimes called a "separation of concerns". For example, there should be no need for *MULT* to be concerned with the internal workings of *ADD*. It needs only to gain access to the result of each addition. It would therefore be appropriate if *MULT* simply copied the two numbers to be added onto a particular tape, then passed control to *ADD*. The latter machine would then leave the result in an agreed format (perhaps on another tape), where it would be collected by *MULT* ready to set up the next call to *ADD*, and so on.

That a multi-tape TM can be replaced by an equivalent single-tape construction means that we can design a TM that possesses any number of tapes, in the knowledge that our machine could be converted into a single-tape machine. We took advantage of this, of course, in the specification of *UTM*, earlier in this chapter.

10.8 The Linguistic Implications of the Equivalence of Non-deterministic and Deterministic TMs

We began by showing that from any non-deterministic TM we could construct an equivalent 4-tape deterministic TM. We then showed that from any multi-tape deterministic machine we could construct an *equivalent* deterministic TM. We have thus shown the following:

For any non-deterministic TM there is an equivalent deterministic TM.

Moreover, in Chapter 7 we showed that the non-deterministic TM is the corresponding abstract machine for the *type* 0 languages (in terms of the ability to *accept*, but not necessarily to *decide* those languages, as we shall see in Chapter 11).

We can now say:

The deterministic TM is the recogniser for the phrase structure languages, in general.

This is so because:

Any phrase structure grammar can be converted into a non-deterministic TM, which can then be converted into a deterministic TM (via the conversion of the non-deterministic TM into an equivalent 4-tape deterministic TM, and *its* conversion into an equivalent deterministic single-tape TM).

EXERCISES

For exercises marked "†", solutions, partial solutions, or hints to get you started, appear in "Solutions to Selected Exercises" at the end of the book.

10.1. Write a program representing a general Turing machine simulator. Your machine should accept as its input the quintuples of a Turing machine, and then simulate the behaviour of the TM represented by the quintuples on various input tapes. In particular, you should run your simulator for various TMs found in this book, and the TMs specified in exercise 2 of Chapter 7. *Hint: the advice given for exercise 6 of Chapter 5, in "Solutions to Selected Exercises" is also very relevant to this exercise.*

10.2. Define a TM to generate $\{a, b, c\}^+$, as outlined earlier in the chapter. Then convince yourself that for any alphabet, A, A^+ is similarly *enumerable*, i.e., we can define a TM to print out its elements in a systematic way.

10.3.† Sketch out the part of the single-tape machine, S, that deals with the sextuple $(1, a, A, R, 2, 2)$ of the three-tape machine, M, described earlier in the chapter.

11

Computability, Solvability and the Halting Problem

11.1 Overview

This chapter investigates the limitations of the TM. The limitations of the TM are, by Turing's thesis (introduced in the last chapter), also the limitations of the digital computer.

We begin by formalising the definition of various terminology used hitherto in this book, in particular, we consider the following concepts of computability:

- functions

- problems

- the relationship between solvability and decidability.

A key part of the chapter investigates one of the key results of computer science, the unsolvability of the *halting* problem, and its theoretical and practical implications. These implications include:

- the inability of any program to determine in general if any program will terminate given a particular assignment of input values

- the existence of decidable, semidecidable and undecidable languages.

The chapter concludes with a specification of the (Turing) computable properties of the languages of the Chomsky hierarchy (from Part 1 of the book).

A.P. Parkes, *A Concise Introduction to Languages and Machines*,
DOI: 10.1007/978-1-84800-121-3_11, © Springer-Verlag London Limited 2008

11.2 The Relationship Between Functions, Problems, Solvability and Decidability

We have said much in the preceding chapters about the power of the TM. We spend much of the remainder of this chapter investigating its *limitations*. In Part 2 of this book there have seen many references to *functions* and *solving problems*. However, no clear definition of these terms has been given. We are now familiar enough with TMs and Turing's thesis to be able to provide these definitions.

11.2.1   Functions and Computability

A *function* associates input values with output values in such a way that for each particular arrangement of input values there is but one particular arrangement of output values. Consider the arithmetic *add* (+) operator. This is a *function* because for any pair of numbers, x and y, the result of $x + y$ is always associated with a number z, which represents the value obtained by adding x and y. In the light of observations made in the preceding two chapters, it should be appreciated that, since any TM can be represented in binary form, it suffices to define the values (input and output) of functions in terms of *numbers*. This applies equally to the language processing machines considered in this book as to those that carry out numerically-oriented operations.

We know that we can represent any of our abstract machines for language *recognition* (*finite state recognisers, pushdown recognisers*) as TMs (see Chapter 7), then represent the resulting TM in such a way that its operation is represented entirely as the manipulation of binary codes. Thus, the notion of *function* is wide-ranging, covering associations between *strings* and *truth values* ("true" if the string is a sentence of a given language, "false" if it is not), numbers and other numbers (*MULT, DIV*, etc.), numbers and truth values (*COMPARE*), and so on.

Calling a function an *association* between input and output values represents a very high level view of a function. We are usually also interested in *how* the association between the input and output values is made, i.e. the *rules* by which we derive a particular arrangement of output values from a particular arrangement of input values. More specifically, we are interested in characteristics of the *program* that embodies the "rules" and produces the output values when presented with the input values.

The last chapter discussed the relationship between TMs and *effective procedures*. We can now say that the latter are those procedures that are:

- deterministic

- finite in time and resources (space)

- mechanical, and

- represented in numeric terms.

It has been argued above that such effective procedures are precisely those that can be carried out by TMs, and, in the light of the previous chapter, we can say by one TM, which we call the *universal TM (UTM)*.

Having a clearer idea of the notions "function" and "effective procedure" enables us to define what we call the *computable functions*, in Table 11.1.

So, Turing's thesis asserts that the *effectively computable functions* and the *TM-computable functions* are the same.

11.2.2   Problems and Solvability

Associated with any function is a *problem*. In abstract terms, the problem associated with a function is a *question*, which asks "what is the result of the function for these particular input values?" For the "+" function, then, an associated problem is "what is $3 + 4$?" It was suggested above that the sentence recognition task for a given language could be regarded as a function associating strings with truth values. The associated problem then asks, for a given string, "what is the truth value associated with this string in terms of the given language?" This is more usually expressed: "is this string a sentence of the language or not?"

A problem is *solvable* if the associated function is computable. We know now that "computable" is the same as saying "TM-computable". So, a problem is solvable if it can be coded in such a way that a TM could compute a solution to it in a finite number of transitions, or a program run on an unlimited storage machine could do the same.

A problem is *totally solvable* if the associated function can be computed by some TM that produces a correct result for every possible assignment of valid input values. An example of this is the "+" operator. We could define a TM, such as the *ADD* machine, that produces the correct result for any pair of input values, by adding the numbers and leaving the result on its tape. Another totally solvable

Table 11.1 "Computable functions" and "TM-computable functions".

The *computable functions* are those functions for which there is an *effective procedure* for deriving the correct output values for each and every assignment of appropriate input values. They can therefore be called the *effectively computable functions*.

A *TM-computable function* is a function that can be computed by some TM.

problem is the *membership problem* for the context sensitive language $\{a^i b^i c^i : i \geq 1\}$. In Chapter 7, we defined a TM that printed T for any string in $\{a, b, c\}^*$ that was a member of the language, and F for any such string which was not. The membership problem for $\{a^i b^i c^i : i \geq 1\}$ is thus *totally solvable*.

A problem is *partially solvable* if the associated function can be computed by some TM that produces a correct result for each possible assignment of input values for which it halts. However, for some input values, the TM will not be able to produce any result. The critical thing is that on any input values for which the TM halts, the result it produces is correct. This concept is best considered in the context of language recognition, and you may already have appreciated the similarity between the concept of *solvability* and the concepts of *deciding* and *accepting* languages discussed in Chapter 7. This similarity is not accidental and we return to it later in the chapter. First, in the next section we consider one of the most important partially solvable problems of computer science.

A problem is *totally unsolvable* if the associated function cannot be computed by any TM. It is worth mentioning at this point that in discussing the concepts of solvability we never exceed the characteristics of effective procedures as defined above: we are still considering deterministic, finite, mechanical, numeric processes. For an abstract, simply stated, yet totally unsolvable problem consider this:

> Define a TM that has as its input a non-packed tape (i.e. an arbitrary number of blank squares can appear between any two marked squares) on which there are as, bs and cs, in any order. The machine should halt and print T if the leftmost marked symbol on its tape is a, and should halt and print F otherwise. The read/write head is initially located on the rightmost marked tape square.

The above machine cannot be defined. Suppose we design a TM to address the above problem that moves left on its tape until it reaches an a, then prints T and halts. Does this mean the problem is *partially* solvable? The answer to this is "no", since partial solvability means that the TM must produce a correct result *whenever* it halts, and our "cheating" TM would not (though it would accidentally be correct some of the time). A TM to even partially solve the above problem cannot be designed. I leave it to you to convince yourself of this.

11.2.3 Decision Problems and Decidability

Problems that ask a question expecting a yes/no answer are sometimes known as *decision problems*. In seeking to solve a decision problem we look for the existence of a program (TM) to resolve each of a set of questions, where the formulation of each question depends on the assignment of specific values to certain parameters. Examples are:

- given two regular grammars, R_1 and R_2, are they equivalent?

- given a number, N, is N prime?

- given a TM, M, and a tape T, will M halt eventually when started on the tape T?

A decision problem is *totally solvable* if we can construct a TM that will provide us with the correct "yes/no" response for each and every possible instantiation of the parameters. Thus, for example, the second decision problem above is totally solvable only if we can construct a TM that decides, for any number, if that number is prime or not. We could do this if we wished, so the second problem is totally solvable.

Of the other two problems, only one is totally solvable, and this is the first. In Chapter 4, we saw what we now know can be called *effective procedures* for converting any regular grammars into non-deterministic FSRs, for converting non-deterministic FSRs into deterministic FSRs, and for converting deterministic FSRs into their minimal form. If two regular grammars are equivalent, their associated minimal machines will be exactly the same, if the states of one minimal machine are renamed to coincide with those of the other. If the states of one of the minimal machines cannot be renamed to make the two minimal machines exactly the same, then the two grammars are not equivalent. We could design a (complicated!) TM to do all of this, and so the equivalence problem for regular grammars is *totally solvable*.

11.3 The Halting Problem

Let us look again at the third of the three *decision problems* presented in the immediately preceding section:

- given a TM, M, and a tape T, will M halt eventually when started on the tape T?

This problem is known as the *halting problem* and was briefly alluded to in the discussion of the power of the FST, at the end of Chapter 8. All of us who have written programs have experienced the practical manifestation of this problem. We run our program, and it seems to be taking an overly long time over its computation. We decide it is stuck in an "infinite loop", and interrupt the execution of the program. We have usually made some error in the design or coding of a loop termination condition. Even commercially supplied programs can sometimes suffer from this phenomenon. Some unexpected combination of values leads to a non-terminating repetition of a sequence of instructions. It would be useful if someone could design a program that examined other programs to see if they would terminate for all of their permissible input values. Such a program would form a useful part of the compiler, which could then give out an error message to the effect that a given program would not terminate for certain inputs.

As we have established, a computer can never be more powerful than a TM, so if the halting problem, as represented in TM terms, is unsolvable, then by extension it is unsolvable even by a computer with infinite memory. We see next that the halting problem is indeed unsolvable in the general sense, though it is *partially solvable*, which is where we begin the investigation.

11.3.1 UTM_H Partially Solves the Halting Problem

Consider again our *universal Turing machine* (UTM), as described in Chapter 10. UTM simulates the computation of any machine, M, on any tape T. From the construction of UTM we can see (as also pointed out in Chapter 10), that this simulation is so faithful to M as to enter an indefinite loop if that is what M would do when processing the tape in question. However, one thing is certain: if M would have reached its halt state on tape T, then UTM will also reach its halt state given the coded version of M and the coded version of T as *its* input. We would therefore simply have to modify the UTM procedure (Table 10.6) so that it outputs a "1" and halts as soon as it finds the code "00" on tape 3 (indicating that the simulated machine M has reached its halt state).

Let us call the new version of UTM, described immediately above, UTM_H. As we have defined it, UTM_H *partially* solves the halting problem. Given any machine M and tape T, appropriately coded, UTM_H will write a 1 (on tape 2) and halt if M would have reached its halt state when presented with the tape T as its input. If M would have stopped in an auxiliary state, because no transition was possible, then UTM_H will not reach its halt state. We could further amend UTM_H to output a 0, then move to its halt state, if no applicable quintuple of M's could be found, and therefore M would have halted in a non-halt state. However, if M would loop infinitely when given tape T, then UTM_H, as defined, has no alternative but to do the same.

It is clear that the critical situations for the halting problem are those in which M loops infinitely. If, in addition to our amendments specified above, we could show how to amend UTM_H so that it halts and writes a 0 when M *would* have looped infinitely on tape T, we would be showing that the halting problem is *totally* solvable.

This last amendment is difficult to imagine, since various conditions can result in infinite loops, and a "loop" in a TM could actually comprise numerous states, as a loop in a program may encompass numerous statements. At this point, then, it seems pertinent to try a different approach.

11.3.2 *Reductio ad Absurdum* Applied to the Halting Problem

Let us assume that we have already amended UTM_H so that it deals with both the halting *and* non-halting situations. This new machine, which we will simply call H, is specified schematically in Figure 11.1.

Figure 11.1 defines H as a machine that prints Y and N rather than 1 or 0. Note that the two arcs entering H's halt state are shown as having the blank as the current symbol. We assume that at that point H has erased everything from its

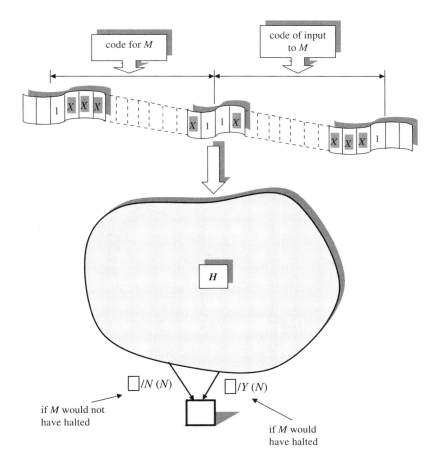

Figure 11.1 The Turing machine, H, is assumed to solve the *halting problem*. Its input is a Turing machine M, and an input tape to M, coded as binary numbers, as described in Chapter 10.

tape, and would thus write Y or N on a blank square before terminating, leaving the read/write head pointing at the result.

So, H is assumed to:

- print Y and halt if M would halt on the input tape T, or

- print N and halt if M would not halt on the tape T.

In other words, H is assumed to *totally solve* the halting problem.

Next, we make what appears to be a strange amendment to H, to create a new machine called H_1, depicted in Figure 11.2.

Figure 11.2 The Turing machine, H of Figure 11.1, is amended to produce the Turing machine H_1 that appears to solve the "*not halting*" *problem*.

In order to produce H_1, we would need to make straightforward amendments to H's behaviour in its halting situations. H_1 would:

- print Y and halt if M would not halt on the input tape T, or

- loop infinitely if M would halt on the tape T.

You should be able to see that, if H existed, it would be straightforward to make the necessary amendments to it to produce H_1.

The next thing we do probably seems even stranger: we modify H_1 to produce H_2, depicted in Figure 11.3.

H_2 copies the code for the machine M, then presents the two copies to H_1. Our scheme codes both machine and tape as sequences beginning and ending with a 1, between these 1s being simply sequences of 0s separated by 1s, so a coded machine could clearly be interpreted as a coded tape.

Thus, instead of simulating the behaviour of M on its input tape, H_2 now simulates the behaviour of the machine M on a description of M itself (!). H_2's result in this situation will then be to:

- print Y and halt if M would not halt on a description of itself, or

- loop infinitely if M would halt on a description of itself.

The next step is the strangest of all. We present H_2 with the input specified in Figure 11.4.

Figure 11.4 shows us that H_2 has been presented with the binary code for itself to represent the machine, M. H_2 will first copy the code out again, then present the two copies to H_1. To H_1, the left-hand copy of the code for H_2 represents a coded TM, the right-hand copy represents the coded input to that TM. Thus, H_2's overall effect is to simulate its own behaviour on an input tape that is a coded version of itself. H_2 uses only the alphabet $\{0, 1, Y, N\}$, as it was derived from H_1, which also uses the same alphabet. Therefore, H_2 could be applied to a description of itself in this way.

11.3.3   The halting problem shown to be unsolvable

Now we ask:

- Does H_2, configured according to Figure 11.4, halt?

    Well, suppose H_2 *does* halt:

- H_2 halts and prints Y if M would not halt on a description of itself.

    However, in the situation here M is H_2, so we have:

- H_2 halts and prints Y if H_2 would not halt on a description of itself.

Figure 11.3 H_2 makes H_1 (of Figure 11.2) apply the machine M to the code for M.

To paraphrase:

- H_2 halts, given a description of itself as its input, if it does not halt, given a description of itself as its input.

    This is clearly nonsense, so we must conclude that H_2 does not halt. However:

- H_2 loops infinitely if M would halt on a description of itself.

Figure 11.4 H_2, of Figure 11.3, is given its own code as input.

Again, in our situation here, M is H_2, so we have:

- H_2 loops infinitely if H_2 would halt on a description of itself.

  To paraphrase once more:

- H_2 does not halt, given a description of itself as its input, if it halts, given a description of itself as its input.

We have devised a machine that halts if it does not halt, and does not halt if it halts. Where was our error? Every change we have made to our machines, from H onwards, requires the addition of a few obvious quintuples. We are forced to go right back and reject our original assumption, as it is obviously this assumption itself that is at fault. The machine, H, which solves the halting problem, cannot exist. The halting problem is an unsolvable problem.

However, to reiterate, the halting problem is *partially solvable*. It is solvable (e.g. by UTM_H, as described above) in precisely those cases when M *would* halt.

Again, it needs to be stated that we are unable to reject the above argument on the basis of the indirection involved in manipulating codes that represent the machines and their tapes. We *cannot*, as computer users, refuse to believe in the representational power of codes. To do so is to denounce as meaningless any activity carried out by real computers.

11.3.4   Some Implications of the Unsolvability of the Halting Problem

The halting problem is one of the most significant results of mathematics. The general unsolvability of the halting problem indicates the ultimate futility of defining effective criteria for determining effectiveness itself. An *algorithm* is an effective procedure that terminates for all of its input values. There can therefore be no *algorithm* to decide, in the general case, whether or not any procedure *is* an algorithm. The unsolvability of the halting problem implies that, in general, we cannot determine by algorithmic means whether or not some well-defined process will terminate for all of its values.

The halting problem also has practical implications for the enterprise of proving the correctness of programs. A program is said to be *partially* correct if whenever it terminates it produces a correct result. A program is said to be *totally* correct if it is partially correct *and* it terminates for each and every possible input value. The goal of proving programs totally correct is thus one that cannot be attained by algorithmic means, as to be able to do so would require a solution to the halting problem.

Many unsolvable problems of computer science reduce to the halting problem. That there is no computational device more powerful than the standard TM implies that effective procedures for analysing the behaviour of other effective procedures are doomed ultimately to simulating those procedures, at the expense of being caught in the never-ending loops of the simulated process, if such loops arise.

As an example of a problem that reduces to the halting problem, consider the *equivalence* problem for TMs, or for programs. Two TMs, M_1 and M_2, are equivalent if for each and every input value, they both yield the same result. Now assume that E is a TM that solves the equivalence problem for any two TMs. Suppose, for argument's sake, that E simulates the behaviour of each machine. It performs the task specified by the next applicable quintuple from M_1, followed by the next applicable quintuple from M_2, on coded representations of an equivalent input tape for each machine. Suppose that the M_1 simulation reaches its halt state first, having produced a result. E must now continue with the M_2 computation until that ends in a result. But what happens if M_2 would not terminate on the tape represented? In order to output F (i.e. the two machines are not equivalent), E would need to determine that M_2 was not going to halt. *E would need to solve the halting problem.*

The halting problem also manifests itself in many decision problems associated with programming. Some of these are:

* will a given program terminate or not? (the halting problem)

* do two programs yield the same answers for all inputs?(this is essentially the TM equivalence problem that is shown to be unsolvable immediately above)

* will a given program write a given value in a given memory location?

* will a given machine word be used as data or as an instruction?

* what is the shortest program equivalent to a given program?

* will a particular instruction ever be executed?

The exercises ask you to represent the above *decision problems* of programming as TM decision problems that can be proved to lack a general solution. In all cases, the method of proof is the same: we show that if the given problem was solvable, i.e., there was a TM to solve it, then that TM would, as part of its activity, have to solve the halting problem.

The proof of the halting problem includes what seems like a peculiar form of *self-reference.* A formal system of sufficient representational power is capable of being used to represent itself. If it is capable of representing itself then it can be turned in on itself, so to speak, to lead to peculiar contradictions. Turing's results are analogous to similar results in the domain of logic established by the great mathematician Kurt Gödel. Here we simplify things greatly, but essentially Gödel showed that a system of logic that is capable of generating all true statements (what is called *completeness*) can be used to generate a statement P, which, in essence refers to itself, saying "P is false".[1] Such a statement can be neither true nor false and is therefore inconsistent (if it is true, then it is false, and *vice versa*). In a way, such a statement says "I am false". In a similar way, our machine H_2,

---

[1] Rather like the person who says: "Everything I say is a lie".

when given its own code as input, says: "I halt if I do not halt, and I do not halt if I halt"!

The general unsolvability of the halting problem and Gödel's findings with respect to completeness leave computer science in a curious position. A machine powerful enough to compute everything we may ask of it is powerful enough to deal with representations of itself. This leads to *inconsistencies*. If we reduce the power of the machine so that inconsistencies cannot occur, then our machines are *incomplete* and will not be able to compute everything that we may reasonably expect of them.

It is worth briefly discussing some wider implications of the halting problem. The inability of algorithmic processes to solve the halting problem does *not* mean that machines are *necessarily* less intelligent that humans. You may think that *we* can look at the texts of our programs and predict that this or that particular loop will not terminate, as we often do when writing programs. This does not mean we can solve the halting problem *in general*, but only in certain restricted situations. There must come a point at which the size and complexity of a program becomes such that we are forced to "execute" (e.g. on paper) the program. There is no reason to disbelieve that, for the restricted cases in which we *can* determine that a program would not halt, we could also write a program (and thus construct a TM) to do the same. The general halting problem embodies a level of complexity that is beyond the capabilities of machines *and* people.

11.4 Computable Languages

In Chapter 7 we saw how from any phrase structure grammar we could construct a non-deterministic TM to *accept* the language generated by that grammar. In Chapter 10 we saw how any *non-deterministic TM* could be converted into an equivalent *deterministic TM*. On this basis, we concluded that the TM is the recogniser for the phrase structure languages in general. The phrase structure languages include all of types 0, 1, 2 and 3 of the *Chomsky hierarchy*. However, as the earlier chapters of this book have shown, the TM is a more powerful computational device than we actually *need* for the more restricted types (context free and regular). Nevertheless, the phrase structure languages are *computable* languages, using the term "computable", as we do now, to mean *TM-computable*. A computable language, then, is a formal language that can be *accepted* by some TM.

A question that arises, then, is "are there *formal languages* that are not computable?" In the next section, we shall encounter a language that is not computable, in that there can be no TM to accept the language. This, in turn, means that there can be no phrase structure grammar to generate that language (if there was we could simply follow the rules in Chapter 7 and construct from the grammar a TM recogniser for the language). This result will enable us to refine our hierarchy of languages and abstract machines, and fully define the relationship between languages and machines.

With respect to formal languages, the analogous concept to *solvability* is called *decidability*.

A language is *decidable* if some TM can take any string of the appropriate symbols and determine whether *or not* that string is a sentence of the language. An alternative way of putting this is to say that the *membership problem* for that language is totally solvable.

A language is *acceptable*, which we might also call *computable*, if some TM can determine if any string that *is* a sentence of the language is indeed a sentence, but for strings that are not sentences we cannot necessarily make any statement about that string at all. In this case, the membership problem for the language is *partially* solvable.

At the end of this chapter, we summarise the decidability properties of the various types of formal language we have encountered so far in this book. For now we return to the question asked above: *are there formal languages that are not computable?* If a language is non-computable, then it is not acceptable, and by Turing's thesis there is no algorithmic process that can determine, for *any* appropriate string, if that string is in the language.

11.4.1   An Unacceptable (non-Computable) Language

We now establish that there is indeed a non-computable formal language. As you will see, the "proof" of this is inextricably bound up with the halting problem, and also takes advantage of our TM coding scheme from the previous chapter.

Consider the following set X:

$$X = \{1x1 \ : \ x \in \{0,1\}^+\}.$$

X is the set of all strings starting and ending in 1, with an arbitrary non-empty string of 0s and 1s in between. Now each string in this set either can, or cannot, represent the code of a TM. By considering a slight adaptation of the TM (sketched out in Chapter 10) that generates all strings in $\{a, b, c\}^+$ as part of its simulation of a non-deterministic machine, we know that the strings of this set could be generated systematically.

Now consider a subset Y of X defined as follows:

$$Y = \{x \ : \ x \in X \text{ and } x \text{ is the code of a TM with alphabet } \{0,1\}\}.$$

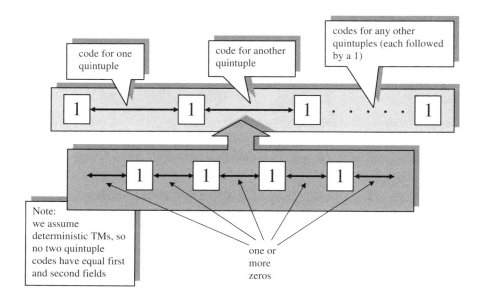

Figure 11.5 The form of a valid binary code to represent a Turing machine, according to the coding scheme of Chapter 10.

It would be straightforward to define a TM, or write a program that, given any string from X, told us whether or not that string was a valid TM code, i.e., of the format shown in Figure 11.5.

Note from Figure 11.5 that we are only interested in codes that represent *deterministic* machines, i.e. machines that have no more than one *quintuple* with the same "current state" and "current symbol". We are also interested only in binary TMs, but we have already argued that all TMs can be reduced to binary form (cf. *UTM*). This means that our "current symbol" and "write symbol" quintuple fields (i.e. the second and third fields of our quintuple representations) will not be allowed to contain more than two 0 s.

The set X can be generated systematically, and we can check for each string whether or not it represents a code for a TM. Therefore, we can see that there is an *effective procedure* for generating the set Y. Note that we are only interested in whether the codes are in the correct form to represent some TM, not in what the TM does. Most of the TM codes in Y would represent machines that do nothing of interest, or, indeed, nothing at all.

Given any code from the set Y we could recover the original machine. We can assume, for argument's sake, that for the alphabet, the code 0 represents 0, 00 represents 1, and all other codes are as we defined them in Chapter 10. If y is a string in the set Y, we will call the machine whose code is y, M_y.

Now consider the following set, Z:

$$Z = \{x \ : \ x \in Y \ and \ M_x \ does \ not \ accept \ x\}.$$

Every string in Y represents a coded TM with alphabet $\{0, 1\}$, and thus every string in Y can be presented as input to the machine whose code is that string. Thus, Z is the set of all binary alphabet machines that do not accept the string represented by their own code.

Now, we ask:

- is Z an *acceptable* language?

Suppose that it is. In this case, some TM, say M_z, is the acceptor for Z. We can assume that M_z uses only a binary alphabet. Thus, M_z's code will be in the set Y. Let's call M_z's code z.

We now ask,

- is the string z accepted by M_z?

Suppose that it is. Then z is not in Z, since Z contains only strings that are not accepted by the machine whose code they represent. It therefore seems that if z is accepted by M_z, i.e. is in the language accepted by M_z, then it isn't in the set Z. However, this is nonsense, because M_z is the acceptor for Z, and so any string it accepts must be in Z.

We therefore reject our assumption in favour of its negation. We suppose that *the string z is not accepted by M_z*. M_z is the acceptor for Z, and so any string M_z does not accept is not in Z. But this cannot be correct, either! If M_z does not accept the string z, then z is one of the strings in Z, as it is the code for a machine (M_z) that does not accept its own code.

It seems that the string z is not in the set Z if it is in the set Z, and is in the set Z if it is not! The TM, M_z cannot exist, and we must conclude that the language Z is not acceptable. We have found an easily defined formal language taken from a two-symbol alphabet that is not computable. Such a language is sometimes called a *totally undecidable* language. We could generalise the above proof to apply to alphabets of any number of symbols, in effect showing that there are an infinite number of such totally undecidable languages.

11.4.2   An Acceptable, But Undecidable, Language

One problem remains to be addressed, before the next section presents an overall summary of the relationship between formal languages and automata. We would like to know if there are there are computable languages that are necessarily acceptable, but not decidable. For reasons that will be presented in the next section, any such languages must be *properly* type 0 in the Chomsky hierarchy, since the more restricted types of language in the hierarchy are *decidable*.

Here is a computable language that is acceptable, but not decidable. Given the sets X and Y, as specified above, we define the set W, as follows:

$$W = \{x \ : \ x \in Y \ and \ UTM_H \ halts \ when \ it \ applies \ M_x \ to \ x\}.^2$$

The set W is the set of codes of TMs that halt when given their own code as input. W is an *acceptable* language. W is not a *decidable* language because a TM that could tell us which valid TM codes were not in W (i.e. represent the code for machines that do not halt when given their own code as input) would have to solve the halting problem.

Languages, like W, that are acceptable but not decidable are sometimes called *semidecidable* languages.

11.5 Languages and Machines

As implied above, associated with any language is what is called the membership problem for that language. Given a grammar G and a string s, the membership problem says: "is s a sentence of the language $L(G)$?" From our discussion of problems and solvability earlier in this chapter, it should be clear that the membership problem for a language is *solvable* if the associated membership *function* for that language is (TM-)computable.

For all the phrase structure languages, the associated membership function is computable. All phrase structure languages are computable (acceptable) languages. However, we have seen that some type 0 languages are *acceptable*, but not *decidable*. We shall see that the other types (1, 2 and 3) are all decidable.

Table 11.2 summarises the decidability properties of the *regular*, *context free*, *context sensitive* and *unrestricted* languages. In fact, since we would like to show only that types 1, 2 and 3 are *decidable*, we can specify the same (very inefficient!) method for all three types, assuming some initial minor amendments to the type 2 and 3 grammars.

---

[2] UTM_H was introduced earlier in the chapter. It halts and prints 1 when the simulated machine would have halted on the given tape, and thus *partially* solves the halting problem.

Table 11.2 Decidability properties for the languages of the Chomsky hierarchy.

Grammar type	General decidability	Justification
regular (type 3)	decidable	As for context free grammars.
context free (type 2)	decidable	If necessary, remove ε from grammar using the method described in Chapter 5, then continue as for context sensitive grammars.
context sensitive (type 1)	decidable	All productions are of the form $x \rightarrow y$ where $|x| \leq |y|$. Thus, each step in a derivation either increases or leaves unchanged the length of the sentential form.
		Construct a non-deterministic TM to generate in parallel all sentential forms that are not greater in length than the target string s. If s is one of the generated strings, print T and halt, if it is not, print F and halt.
		Convert the non-deterministic TM into an equivalent deterministic TM (see Chapter 10).
unrestricted (type 0)	semi-decidable	Some productions may be of the form $x \rightarrow y$ where $|x| > |y|$. The sentential form may thus lengthen and shorten during a derivation. There may therefore be no obvious point at which to give up if we do not manage to generate s, the target string.
		However, we know from Chapter 7 that all type 0 languages are *acceptable*.

In terms of formal languages, as presented in Part 1 of this book, we described each class of language in the Chomsky hierarchy, working from the regular languages up through the hierarchy. We simultaneously worked upwards from the finite state recogniser (FSR) to the TM, by adding extra functionality to one type of abstract machine to yield a more powerful type of machine. For example, we showed how a machine that is essentially an FSR with a stack (a pushdown recogniser) was more powerful than an FSR. We finally developed the Turing machine and showed that we need go no further. There is no more powerful device for computation or language recognition than the TM.

We might, as an alternative, have approached the subject by first presenting the TM, and then discussing how the TM actually gives us more power than we need for the more restricted languages and computations (such as *adding* and *subtracting*, which we saw being performed by FSTs in Chapter 8). Then, the abstract machines for the more restricted languages could have been presented as TMs that were restricted in some appropriate way. Figure 11.6 presents this alternative but equivalent view of computable languages and abstract machines.

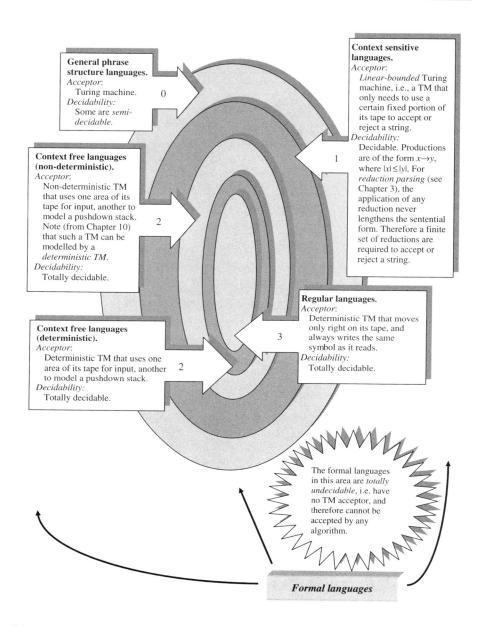

General phrase structure languages.
Acceptor:
Turing machine.
Decidability:
Some are *semi-decidable*.

0

Context sensitive languages.
Acceptor:
Linear-bounded Turing machine, i.e., a TM that only needs to use a certain fixed portion of its tape to accept or reject a string.
Decidability:
Decidable. Productions are of the form $x \rightarrow y$, where $|x| \leq |y|$. For *reduction parsing* (see Chapter 3), the application of any reduction never lengthens the sentential form. Therefore a finite set of reductions are required to accept or reject a string.

1

Context free languages (non-deterministic).
Acceptor:
Non-deterministic TM that uses one area of its tape for input, another to model a pushdown stack. Note (from Chapter 10) that such a TM can be modelled by a *deterministic TM*.
Decidability:
Totally decidable.

2

Regular languages.
Acceptor:
Deterministic TM that moves only right on its tape, and always writes the same symbol as it reads.
Decidability:
Totally decidable.

3

Context free languages (deterministic).
Acceptor:
Deterministic TM that uses one area of its tape for input, another to model a pushdown stack.
Decidability:
Totally decidable.

2

The formal languages in this area are *totally undecidable*, i.e. have no TM acceptor, and therefore cannot be accepted by any algorithm.

Formal languages

Figure 11.6 Languages and machines (the numbers denote the type of language in the *Chomsky hierarchy* — see Chapter 2).

EXERCISES

For exercises marked "†", solutions, partial solutions, or hints to get you started appear in "Solutions to Selected Exercises" at the end of the book.

    11.1.[†] The printing problem asks if a TM, M, will print out a given symbol, s, in M's alphabet, when started on a specified input tape. Prove that the printing problem is, in general, unsolvable.
        Hint: this is a problem that reduces to the halting problem, as discussed earlier.

    11.2. Show that the following (all taken from this chapter) are all unsolvable problems. In each case first transform the statement of the problem so that it is couched in terms of Turing machines rather than programs. Then show that in order to solve the given problem, a TM would have to solve the halting problem.

        (a) will a given program write a given value in a given memory location?

        (b) will a given machine word be used as data or as an instruction?

        (c) what is the shortest program equivalent to a given program?

        (d) will a particular instruction ever be executed?

    11.3. Express the totally unsolvable problem in section 11.2.2 in terms of programs, rather than TMs.
        Hint: consider a non-terminating program such as an operating system, for which the input can be regarded as a stream of characters.

12
Dimensions of Computation

12.1 Overview

In previous chapters, we have referred to computational power in terms of the task modelled by our computations. When we said that two processes are *equivalent*, this meant that they both modelled the same task. In this chapter, we refine our notions of computational power, by discussing its three dimensions:

- *function*: *what* is computed

- *space*: *how much storage* the computation requires

- *time*: *how long* the computation takes.

In particular, we introduce a simple model of time into our analysis, in which each transition of a machine is assumed to take one *moment*. This leads us to discuss key concepts relating to both abstract machines and algorithms:

- the implications of parallel processes

- predicting the running time of algorithms in terms of some measure of the "size" of their input, using so-called "big O" notation.

  We discover that algorithms can be classified has having

- linear,

- logarithmic,

A.P. Parkes, *A Concise Introduction to Languages and Machines*,
doi: 10.1007/978-1-84800-121-3_12, © Springer-Verlag London Limited 2008

- polynomial, or

- exponential

running times.

A theme that runs through most of the chapter is an as yet unsolved problem of computer science. In essence, this problem asks whether, in general, parallel processes can be modelled by similarly efficient serial processes.

12.2 Aspects of Computation: Space, Time and Complexity

Earlier in this book, we have seen that the limitations, in terms of computational power, of the *Turing machine* (*TM*) are also the limitations of the modern digital computer. We argued that, simple as it is, the TM is actually *more* powerful than *any* real computer unless we allow for an arbitrary amount of storage to be added to the real computer, as required, during the course of its activity. Even then, the computer would not be *more* powerful than the TM. However, claims about the power of the TM must be appreciated in terms of the TM being an *abstract machine*. In Chapter 8, when discussing the FST, we presented so-called *memory machines*, that "remember" an input symbol for a fixed number of state transitions before that symbol is output. In a sense, by doing this we were using an implicit model of *time*, and the basic assumption of that model was that the transition between two states of a machine took exactly one unit of this time, which we called a *moment*.

Turing's thesis, as discussed in Chapter 10, reflects a *functional* perspective on computation: it focuses on *what* the TM can perform. However, as users and programmers of computers, we are not only interested in the functions our computer can carry out. At least two other fundamental questions concern us, these being:

- how much memory and storage will our task require (*space* requirements)?

  and

- how long will it take for our task to be completed (*time* requirements)?

In other words, the dimensions of computation that occupy us are *function*, *space* and *time*. It is clear then that reference to computational *power*, a term used freely in the preceding chapters, should be qualified in terms of those dimensions of computational power to which reference is being made.

Now, in terms of *space* requirements, our binary TMs of Chapter 9 seem to require an amount of memory (i.e. occupied tape squares) that does not differ significantly from that required by a computer carrying out a similar task on a similar set of data. For example, consider the descriptions of the memory accessing TMs towards the end of Chapter 9. Apart from the binary numbers being processed, the only extra symbols that our machines required were special markers to separate one number from the next. However, these markers were needed because we had implicitly allowed for variable memory "word" (or array element) length. If we had introduced a restriction on word length, as is the case in most memory schemes, we could have dispensed with the need for such markers. However, if we *had* dispensed with the markers between memory items, we would have had to "program in" the knowledge of where the boundaries between items occurred, which in terms of TMs means we would have had to introduce more states to deal with this.

The discussion above seems to imply that we can reduce the number of states in a (well-designed) TM *at the expense of introducing extra alphabet symbols*, and we can reduce the number of alphabet symbols *at the expense of introducing extra states*. In fact, this relationship has been quantified, and it has been shown that

- any TM that has more than two states can be replaced by an equivalent TM that has only two states.

More importantly, from our point of view, it has been shown by the information theorist Claude Shannon, that

- any TM that uses an alphabet of more than two symbols can be replaced by an equivalent TM that uses only two symbols.

The fact that any TM can be replaced by an equivalent TM using only a two-symbol (i.e., *binary*) alphabet is of fundamental importance to the practical use of computers. It indicates that any shortfalls in functional power that real computers have are certainly not a consequence of the fact that they only use a two-symbol alphabet. Due to the reciprocal relationship outlined above between the number of states in a TM and the number of alphabet symbols it needs, the product of the number of states and the number of alphabet symbols is often used as a measure of the *complexity* of a TM. There are, however, other aspects of TMs that could be said to relate to their complexity. One of these aspects is non-determinism, which, as we discovered in the previous chapter, we can dispense with (in *functional* terms). However, as we see next, there are implications for the *space* and *time* dimensions of computation if we remove non-determinism from our machines.

12.3 Non-Deterministic TMs Viewed as Parallel Processors

We saw earlier in this chapter that computation is concerned not only with *function*, but also with *space* and *time*. When we make statements about the *equivalence* of Turing machines, as we have in this chapter, we are usually talking about equivalence of *function*. For example, we say that the result of the "non-deterministic to deterministic TM conversion" of Chapter 10 is "equivalent" to the original machine. However, there are different ways of viewing the operation of a non-deterministic machine. One is to imagine that the machine attempts to follow a sequence of transitions until it finds a solution or reaches a "dead end", as none of its available instructions can be applied. In the latter case, we have assumed that it then tries an alternative sequence. When combined with a notion of *time* that is essentially a measure of the number of *transitions* made by the machine, this perspective reflects a *serial* model of computation. Our earlier discussions regarding non-deterministic *finite state recognisers* (Chapter 4) and *pushdown recognisers* (Chapter 5) also reflect this view, when considering concepts such as *backtracking*, for example.

As an alternative to a serial model of execution, we might assume that all possible sequences of transitions to be considered by the non-deterministic TM are investigated in *parallel*. When the machine is ready to start, imagine that a signal arises, analogously to the way in which the internal "clock" governs the state changes of the CPU. On this signal, each and every applicable transition is carried out simultaneously (each transition being allowed to take with it its own copy of the input tape, since each sequence of transitions may result in different configurations of the tape). Figure 12.1 shows an example starting situation with three applicable transitions.

Figure 12.1 is intended to represent the situation that arises on the first "*clock* tick", which we assume results in the execution of the three applicable transitions in parallel (the result of each being shown by the dashed arrows). Note that the new tape configuration for the three cases includes an indication of which state the machine has reached after making the relevant transition. A description of the marked portion of a tape, the location of the read/write head and the current state of a machine at any stage in a TM's computation is sometimes called an *instantaneous description*.

The next clock tick would result in the *simultaneous* execution of:

- each applicable transition from state 2 given the tape configuration on the left of Figure 12.1,

- each applicable transition from state 3 given the tape configuration in the centre of Figure 12.1, and

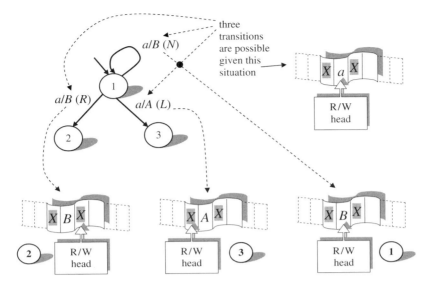

Figure 12.1 The transitions of a non-deterministic Turing machine viewed as parallel processes.

- each applicable transition from state 1 given the tape configuration on the right of Figure 12.1.

Then the next clock tick would result in the simultaneous execution of each applicable transition from each of the situations created at the previous clock tick.

Clearly, if a solution exists the above method will find it. In fact, the first sequence of transitions we find that leads us to a solution will be the shortest such sequence that can be found.[1] Now consider a deterministic 4-tape version of the non-deterministic machine (as described in Chapter 10). This machine executes all possible sequences of quintuples of the non-deterministic machine, but it *always explores shorter sequences first.* Thus, the deterministic 4-tape machine is also bound to discover the shortest sequence of quintuples that will lead to a solution.

---

[1] The shortest solution sequence may not be unique. In this case, our parallel machine will find all of the shortest solutions simultaneously. It will simplify subsequent discussions if we talk as if there is only one shortest solution.

12.3.1   Parallel Computations and Time

Now our model of *time* enters the analysis. If there is a solution that requires the execution of n instructions (i.e. n *quintuples*), the parallel machine will find it in n *moments*. In our example above, if any applicable sequence of transitions leads directly to the halt state from any of the resulting configurations shown, our parallel machine would have taken 2 moments to find the solution. The first transitions would be simultaneously executed in one moment, then all of the applicable transitions from the three resulting configurations would similarly be executed in the next moment.

However, consider how long it takes our *deterministic* version to find a solution that requires the execution of n instructions. The deterministic machine will have tried:

- all single instructions

- all sequences of two instructions

- .

- .

- some or all of the sequences of n instructions.

Suppose the alphabet of quintuple labels (as in the example in Chapter 10), is $\{a,b,c\}$, i.e. we are talking of a 3-quintuple machine. The deterministic machine generates strings of these by generating all strings of length 1, all strings of length 2, and so on. There would be 3 strings of length 1, 9 strings of length 2, and 27 strings of length 3, which are:

> *aaa, aab, aac, aba, abb, abc, aca, acb, acc, baa, bab, bac, bba, bbb, bbc, bca, bcb,*
> *bcc, caa, cab, cac, cba, cbb, cbc, cca, ccb, ccc.*

At each stage, the strings for the next stage were derived by appending each of the three symbols of the alphabet to each of the strings obtained at the previous stage. Thus, for our example here, there would be 81 (i.e. 27×3) strings of length 4, then 243 (81×3) strings of length 5, and so on.

In general, the number of strings of length n from an alphabet of k symbols is k^n (in the example above, where $n = 4$ and $k = 3$ we had 3^4, or 81, strings).

To return to our main discussion, consider a 3-quintuple non-deterministic parallel TM, N. Suppose the shortest solution that this machine could find, given a certain *input configuration*, involves 4 instructions. N will take 4 moments to find it. Now consider the deterministic version, D, constructed using our method described earlier. We will simplify things considerably by considering the execution time of D in terms of the number of quintuples of N it tries. We will assume that it takes one moment to try each quintuple. We will also ignore the fact that a large number of quintuple sequences will be illegal and thus aborted by D before the entire sequence has been applied. It is reasonable to ignore these things, as D has to *generate* each sequence of quintuple labels, even if the corresponding sequence of quintuples subsequently turns out to be inapplicable. This machine first tries:

3 sequences of instructions of length 1 (3 moments)
9 sequences of instructions of length 2 (18 moments)
27 of length 3 (81 moments)
between 1 and 81 sequences of length 4 (between 4 and 324 moments).

Adding all these up, we get $3 + 18 + 81 + 4 = 106$ moments in the best case (i.e. when the first 4-quintuple sequence tried results in a solution). In the worst case (when all of the 4-quintuple sequences have to be tried, the last one resulting in the solution) we get $3 + 18 + 81 + 324 = 426$ moments. In general, given k quintuples in the non-deterministic machine, and a shortest solution requiring n instructions, the parallel TM,N, will take n moments to find the solution, while the corresponding deterministic machine, D, will take somewhere between:

$$(k \times 1) + (k^2 \times 2) + \ldots + (k^{n-1} - 1) + (k^n \times n)$$
$$= k + 2k^2 + \ldots + (n-1)k^{n-1} + nk^n \text{ moments in the worst case,}$$

and

$$(k \times 1) + (k^2 \times 2) + \ldots + (k^{n-1} \times n - 1) + n$$
$$= k + 2k^2 + \ldots + (n-1)k^{n-1} + n \text{ moments in the best case.}$$

Now, as we have already seen, even simple Turing machines, such as those we have seen in this book, consist of many more than three quintuples. Moreover, solving even the simplest problems usually requires the application of a sequence considerably longer than four quintuples. For example, if a 10-quintuple non-deterministic parallel TM (a very simple machine) found the solution in 11 moments (a very short solution), the best the corresponding deterministic TM constructed by our method could hope to do (as predicted by the above formula) is find a solution in 109 876 543 221 moments. To consider the extent of the difference in magnitude here, let us suppose a moment is the same as 1 millionth of a second (a reasonably slow speed for a modern machine to carry out an operation as simple as a Turing machine transition). Then:

- the non-deterministic machine executes the 11 transitions in slightly over $\frac{1}{100,000}$ of a second, while

- its deterministic counterpart takes around 30.5 hours to achieve the same solution!

The above example may seem rather contrived, but consider that a parallel algorithm is, say, updating a complex graphical image, and millions of individual operations are being carried out simultaneously. Such computations would be impossible to carry out in a reasonable time by a serial processor. We would not wish to wait 30.5 hours for an image in an animation to change, for example.

12.4 A Brief Look at an Unsolved Problem of Complexity

The preceding discussion leads us to an as yet unsolved problem of computer science. You may, on reading the discussion, think that the disparity between the non-deterministic TMs and their 4-tape deterministic counterparts is entirely due to the rather simple-minded conversion scheme we have applied. You may think that there is a more "efficient" way to model the behaviour of the non-deterministic machine than by attempting to execute every possible sequence of its quintuples. We have shown that for any non-deterministic TM we can construct a deterministic TM that does the same job. The question that now arises is: can we construct one that does the same job as efficiently? In order to discuss whether this question can be answered, we take a slightly more detailed look at complexity.

12.5 A Beginner's Guide to the "Big O"

As far as we are concerned here, the theory of complexity examines the relationship between the length (number of symbols) of the input to a TM and the number of instructions it executes to complete its computation. The same applies to programs, where the relationship concerned is between the number of input elements and the number of program instructions executed. In considering this relationship, we usually take account only of the components of the program that dominate the computation in terms of time expenditure.

12.5.1   Predicting the Running Time of Algorithms

Suppose we are to sum all of the elements in a two-dimensional array of numbers, then print the result, and we can only access one element of the array at a time.

We *must* visit each and every element of the array to obtain each value to be added in to the running total. The time taken by the program is *dominated* by this process. If the array is "square", say, $n \times n$, then we are going to need n^2 accessing operations. The times taken by the other operations (e.g. adding the element to the running total in the loop, and printing out the result) we can ignore, as they do not change significantly whatever the size of n.

In overall terms, our array summing program can be said to take an amount of time approximately proportional to $n^2 + c$. Even if we had to compute the array subscripts by adding in an offset value each time, this would not change this situation (the computation of the subscripts would be subsumed into c). Obviously, since c is a constant measure of time, the amount of time our algorithm takes is dominated by n^2 (for large values of n). n^2 is called the *dominant expression* in the equation $n^2 + c$. In this case, then, we say our array processing program takes time *in the order of n^2*. This is usually written, using what is called *big O* – "Oh", not nought or zero – notation, as $O(n^2)$. Similarly, summing the elements of a one-dimensional array of length n would be expected to take time $O(n)$. There would be other operations, but they would be assumed to take constant time, and so n would be the dominant expression. In a similar way, summing up the elements of a three-dimensional $n \times n \times n$ array could be assumed to take time $O(n^3)$.

In estimating the running time of algorithms, as we have said above, we ignore everything except for the dominant expression. Table 12.1 shows several running time expressions and their "big O" equivalents.

Table 12.1 Some expressions and their "big O" equivalents. In the Table, $c_1 \dots c_j$ are constants, n is some measure of the size of the input.

Expression	Big O equivalent	Type of running time
c	$O(1)$	Constant
$c_1 + c_2 + \dots + c_j$	$O(1)$	Constant
n	$O(n)$	Linear
cn	$O(n)$	Linear
n^2	$O(n^2)$	Polynomial
cn^2	$O(n^2)$	Polynomial
$c_1 n^{c2}$	$O(n^{c2})$	Polynomial
c^n	$O(c^n)$	Exponential
$n + c_1 + c_2 + \dots + c_j$	$O(n)$	Linear
$n^c + n^{c-1} + n^{c-2} + \dots + n$	$O(n^c)$	Polynomial (if $c > 1$)
$(1/c_1)n^{c2}$	$O(n^{c2})$	Polynomial (if $c_2 > 1$)
$(n^2 + n)/2$	$O(n^2)$	Polynomial
$(n^2 - n)/2$	$O(n^2)$	Polynomial
$\log_2 n$	$O(\log_2 n)$	Logarithmic
$n \times \log_2 n$	$O(n \times \log_2 n)$	Logarithmic

Finally, if it can be demonstrated that an algorithm will execute in the least possible running time for a particular size of input, we describe it as an *optimal* algorithm. A large part of computer science has been concerned with the development of optimal algorithms for various tasks. The following discussion briefly addresses some of the key issues in algorithm analysis. We discuss the main classes of algorithmic running times, these being *linear, logarithmic, polynomial* and *exponential.*

12.5.2   Linear time

If the running time of an algorithm is described by an expression of the form $n + c$, where n is a measure of the size of the input, we say the running time is $O(n)$ – "of the order of n" – and we describe it as a *linear time* algorithm. The "size" of the input depends on the type of input with which we are dealing. The "size" of a string, would be its *length*. The "size" of an array, list or sequence would be the number of elements it contains.

As an example of a linear time algorithm, consider the sequential search of an array of n integers, sorted in ascending order. Table 12.2 shows a Pascal-like program to achieve this task. As stated earlier, the components assumed to require constant running time are ignored. The key part of the algorithm for the purposes of running time estimation is indicated in the table.

If you are alert, you may have noticed that the algorithm in Table 12.2 does not always search the whole of the array (it gives up if it finds the number or it

Table 12.2 *Linear search* of a one-dimensional array, *a*, of n integers, indexed from 1 to n, sorted in ascending order.

{target *is an integer variable that holds the element to be found*}	
done := false	
foundindex := 0	{*If* target *is found,* foundindex *will hold its index number. If not found,* foundindex *will remain at* 0}
i := 1	
while i ≤ n and not(done) do	{*the number of times this loop iterates, as a function of n, the size of the input, is the key element in the algorithm's running time*}
if target = a[i] then	{*found*}
done := true	
foundindex := i	
else	
if target < a[i] then	{*we are not going to find it*}
done := true	
endif	
endif	
i := i + 1	
endwhile	

becomes obvious that the number isn't there). For many algorithms, a single estimate of the running time is insufficient, since properties of the input other than its size can determine their performance. It is therefore sometimes useful to talk of the *best*, *average* and *worst* case running times. The best, worst and average cases for the algorithm in Table 12.2, and their associated running times, are shown in Figure 12.2. You will note that though the average case involves

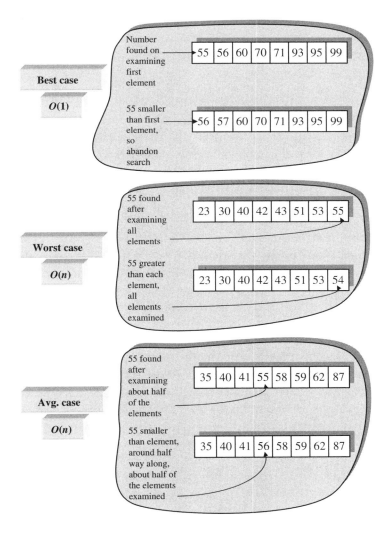

Figure 12.2 The best, worst and average cases for linear search of a sorted array of integers, when the number being searched for is 55.

$n/2$ searches, as far as big O is concerned, this is $O(n)$. We ignore coefficients, and $n/2 = 1/2n$, so the "1/2" is ignored.

Since an abstract machine can embody an algorithm, it is reasonable, given that model of time is based on the number of transitions that are made, to talk of the efficiency of the *machine*. A deterministic finite state recogniser, for example, embodies an $O(n)$ algorithm for accepting or rejecting a string, if n is the length of the string.

12.5.3 Logarithmic Time

Consider again the algorithm of Table 12.1. As stated earlier, it performs a linear search of a sorted array. To appreciate why this algorithm is not optimal, consider a telephone directory as being analogous to a sorted array. If I asked you to look up "Parkes" in the directory, you are hardly likely to begin at the first entry on page 1 and completely peruse, in order, all of the entries on that page before moving to page 2, and so on. You are more likely to open the book somewhere where you think the "Ps" might be. If you are lucky, you will find "Parkes" on that very page. Suppose the page you have reached has names beginning with "S". You then know you've gone too far, so you open the book at some point before the current page, and so on. The process you are following is a very informal version of what is known as *binary search*. We consider this in more detail now.

Binary search searches for a given item within a sorted list (such as an array), as does linear search, discussed above. However, it is much more efficient. Figure 12.3 schematically represents the operation of binary search on a particular example. Binary search is represented algorithmically in Table 12.3.

So what is the running time of binary search? As hinted in Figure 12.3, it turns out to be $O(\log_2 n)$, i.e., the running time is proportional to the binary logarithm of the number of input items. Many students have a problem in grasping logarithmic running times, so we will spend a little time considering why this is so.

Firstly, for those who have forgotten (or who never knew) what logarithms are, let us have a brief revision (or first time) session on "logs". Some of you may be used to base 10 logarithms, so we'll start there. If x is a number, then $\log_{10} x$ is the number y, such that $10^y = x$. For example, $\log_{10} 100 = 2$, since $10^2 = 100$. Similarly, $\log_{10} 1000 = 3$, since $10^3 = 1000$. We can get a reasonable estimate of the $\log_{10}$ of a number by repeatedly dividing the number by 10 until we reach 1 or less. The number of divisions we carry out will be the estimate of the $\log_{10}$ value. Take 1000 for example:

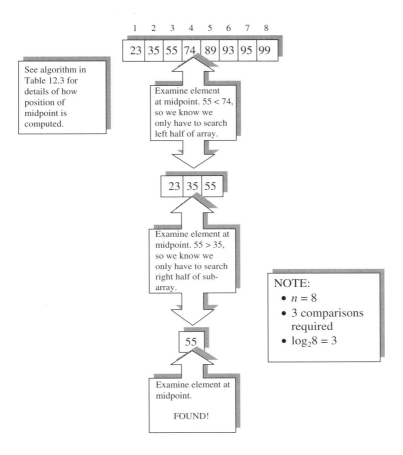

See algorithm in Table 12.3 for details of how position of midpoint is computed.

Examine element at midpoint. 55 < 74, so we know we only have to search left half of array.

Examine element at midpoint. 55 > 35, so we know we only have to search right half of sub-array.

NOTE:
- $n = 8$
- 3 comparisons required
- $\log_2 8 = 3$

Examine element at midpoint.

FOUND!

Figure 12.3 How binary search (Table 12.3) works, when trying to find 55 in a sorted array of integers.

$$1000/10 = 100 - \text{step } 1$$
$$100/10 = 10 - \text{step } 2$$
$$10/10 = 1 - \text{step } 3.$$

We require 3 steps, so $\log_{10} 1000$ is approximately (in fact it is *exactly*) 3. Let us consider a less convenient example. This time we'll try to estimate $\log_{10} 523$:

$$523/10 = 52.3 - \text{step } 1$$
$$52.3/10 = 5.23 - \text{step } 2$$
$$5.23/10 = 0.523 - \text{step } 3.$$

Table 12.3 *Binary search* of a one-dimensional array, *a*, of *n* integers, indexed from 1 to *n*, sorted in ascending order.

{target *is an integer variable that holds the element to be found*}	
low := 1	
high := n	{low *and* high a*re used to hold the lower and upper indexes of the portion of the array currently being searched*}
foundindex := 0	{*If* target *is found,* foundindex *will hold its index number. If not found,* foundindex *will remain at* 0}
done := false	
while low ≤ high and not(done) do	
mid := (low + high) div 2	{*calculate midpoint of current portion of array* (div *is integer division*)}
if target = a[mid] then	{*found*}
foundindex := mid	
done := true	
else	
if target < a[mid] then	{*need to search left half of current portion*}
high := mid – 1	
else	
low := mid + 1	{*need to search right half of current portion*}
endif	
endif	
endwhile	

Here, the result of the final step is *less than* 1. This tells us that $\log_{10}523$ is more than 2 but less than 3 (it is in fact 2.718501689, or thereabouts!). That this approach is a gross simplification does not concern us here; for the purposes of algorithm analysis it is perfectly adequate (and as we shall shortly see, very useful in analysing the running time of binary search).

The same observations apply to logarithms in any number base. If x is a number, $\log_2 x$ is the number y, such that $2^y = x$. For example, $\log_2 64 = 6$, as $2^6 = 64$. We can use the same technique as we used for $\log_{10}$ to estimate the $\log_2$ of a number (but this time we divide by 2). Let us consider $\log_2 10$:

$$10/2 = 5 - \text{step } 1$$

$$5/2 = 2.5 - \text{step } 2$$

$$2.5/2 = 1.25 - \text{step } 3$$

$$1.25/2 = 0.625 - \text{step } 4.$$

So we know that $\log_2 10$ is more than 3 but less than 4 (it's actually around 3.322).

It turns out that our method for estimating the $\log_2$ of a number is very relevant in analysing the running time of binary search. As you can see from Figure 12.3, binary search actually reduces by half the number of items to be searched at each stage. Assume the worst case, which is that the item is found as the last examination of an element in the array is made, or the searched for item is not in the array at all. Then the number of items examined by binary search beginning with an array of n items will closely follow our method for estimating the $\log_2$ of n:

examine an item, split array in half $[n/2 - \text{step 1}]$

examine an item, split half array in half $[(n/2)/2 - \text{step 2}]$

.

.

... until only an array of 1 element remains to be searched. Hence, the number of items in an n-element array that need to be examined by binary search is around $\log_2 n$ in the worst case. All of the other operations in the algorithm are assumed to be constant and, as usual, can be ignored. The running time of binary search is thus $O(\log_2 n)$.

In general, any algorithm that effectively halves the size of the input each time it loops will feature a logarithm in the running time. The importance of logarithmic running time is that it represents a great saving, even over a linear time algorithm. Consider the different in performance between linear search, discussed above, and binary search if the number of items to be searched is 1 000 000. In cases tending towards the worst, linear search may have to examine nearly all of the 1 000 000 items, while binary search would need to examine only around 20.

Binary search is justifiably a famous algorithm. Also among the better known of the logarithmic algorithms is a sorting algorithm known as *quicksort*. Quicksort sorts an array of integers, using the approach schematically represented for an example array in Figure 12.4. As can be seen, the algorithm works by placing one item (the *pivot*) into its correct position, then sorting the two smaller arrays on either side of the pivot using the same (quicksort) procedure. However, quicksort is not just a $\log_2 n$ algorithm, but actually an $O(n \times \log_2 n)$ algorithm, as hinted in Figure 12.4. As we shall see shortly, many established sorting algorithms are $O(n^2)$, and thus $n \times \log_2 n$ is a considerable improvement. An n^2 sorting algorithm would require about 225 000 000 units of time to sort 15 000 items, while quicksort could achieve the same job in 210 000 time units.

12.5.4 Polynomial Time

Programs that take times such as $O(n^2)$ and $O(n^3)$, where n is some measure of the size of the input, are said to take *polynomial time* in terms of the input

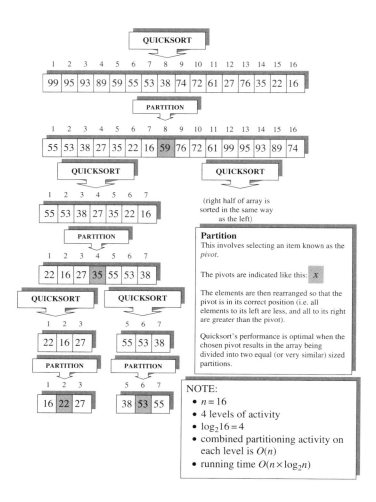

Figure 12.4 How *quicksort* sorts an array of 16 integers into ascending order.

data. Consider an algorithm related to our linear search, above. This time, the algorithm searches for an element in a *two*-dimensional $(n \times n)$ array, sorted as specified in Figure 12.5. This can be searched in an obvious way by the program in Table 12.4. This algorithm has a running time $O(n^2)$ for both the average case (target element somewhere around position $a[n/2, n/2]$ – or search can be abandoned at this point) and worst case (target element at position $a[n, n]$, or not in array at all). The average case is $O(n^2)$ because as far as big O analysis is concerned, $n^2/2$ might just as well be n^2.

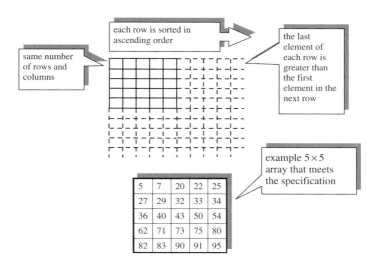

Figure 12.5 A sorted two-dimensional $n \times n$ array – specification and example.

Table 12.4 The obvious, but inefficient, algorithm to search a two-dimensional $n \times n$ array, a, of integers, sorted as specified in Figure 12.5.

{target *is an integer variable that holds the element to be found*}
done := false
foundindexi := 0
foundindexj := 0   {*If* target *is found,* foundindexi *and* foundindexj *hold its index values*}
i := 1   {*i is row index variable*}
while i ≤ n and not(done) do {*outer loop controls row index*}
 j := 1   {*j is column index variable*}
 while j ≤ n and not(done) {*inner loop controls column index*}
  do
  if target = a[i, j] then {*found*}
   done := true
   foundindexi := i
   foundindexj := j
  else
   if target < a[i, j] then {*we are not going to find it*}
    done := true
   endif
  endif
  j := j + 1
 endwhile
 i := i + 1
endwhile

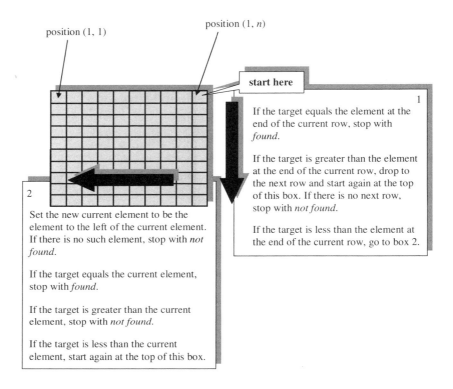

Figure 12.6 A sketch of an $O(n)$ algorithm to search for an element in an $n \times n$ array that meets the specification in Figure 12.5.

However, there is a simple but clever algorithm that can do the task in $O(n)$ time. The algorithm is outlined in Figure 12.6. An exercise asks you to convince yourself that its running time is indeed $O(n)$.

As a final example of an $O(n^2)$ algorithm, let us reopen our discussion of sorting which began with our analysis of *quicksort*, above. We consider the algorithm known as *bubble* (or sometimes *exchange*) sort. This algorithm is interesting partly because it has a best case running time of $O(n)$, though its worst case is $O(n^2)$. The bubble sort algorithm is shown in Table 12.5. Figure 12.7 shows a worst case for the algorithm based on a six-element array. The algorithm gets its name from the way the smaller elements "bubble" their way to the "top".

Considering the bubble sort algorithm (Table 12.5) and the trace of the worst case (Figure 12.7), it is perhaps not obvious that the algorithm is $O(n^2)$. The outer loop iterates $n-1$ times, as pointer i moves up the array from the second to the last

Table 12.5 The *bubble* (or *exchange*) *sort* procedure that sorts an array, *a*, of *n* integers in ascending order. The operation of this algorithm is shown in Figure 12.7.

i := 2	
sorted := false	
while i ≤ n and not(sorted) do	
sorted := true	{*if* sorted *is still true at the end of the outer loop, the array is sorted and the loop terminates*}
for j := n downto i do	{downto *makes the counter* j *go down from* n *to* i *in steps of* 1}
if a[j] < a[j − 1] then	
sorted := false	{*the following three statements swap the elements at positions* j *and* j − 1}
temp := a[j]	
a[j] := a[j − 1]	
a[j − 1] := temp	
endif	
endfor	
i := i + 1	
endwhile	

element. However, for each value of i, the second pointer, j, moves from position n down to position i. The number of values taken on by j then, is not simply n^2, but rather $(n − 1) + (n − 2) + ... + 1$. Let us write this the other way round, i.e.

$$1 + ... + (n − 2) + (n − 1).$$

Now, there is an equation that tells us that

$$1 + 2 + ... + n = (n^2 + n)/2.$$

You may like to verify this general truth for yourself. However in our case, we have a sum up to $n − 1$, rather than n. For $n − 1$, the equation should be expressed as:

$$((n − 1)^2 + (n − 1))/2$$
$$= (n^2 + 1 − 2n + n − 1)/2$$
$$= (n^2 − n)/2.$$

This tells us that, in the worst case for bubble sort, the inner loop goes through $(n^2 − n)/2$ iterations. $(n^2 − n)/2 = 1/2n^2 − 1/2n$. Ignoring the coefficients and the smaller term, this reduces to a big O of $O(n^2)$.

    An exercise asks you to define the best case scenario for bubble sort, and to argue that in the best case it has a running time $O(n)$.

Figure 12.7 An example worst case scenario for *bubble sort* (see Table 12.5), resulting in a running time $O(n^2)$.

We complete the discussion of polynomial time algorithms with a look at an $O(n^3)$ algorithm. The algorithm we consider is known as Warshall's algorithm. The algorithm applies to *directed graphs* (an abstract machine is one form of directed graph). The algorithm computes what is known as the *transitive* closure of a directed graph. It takes advantage of the fact that if there is an arc from one node, A, to another, B, and an arc from node B to node C, there is a path from node A to C, and so on. We shall apply Warshall's algorithm to the problem of seeing if all non-terminals in a regular grammar feature in the derivation of a terminal string. To do this we will use the following grammar, G:

$$S \rightarrow aB \mid aC$$
$$B \rightarrow cB \mid cD$$
$$C \rightarrow cC \mid fD \mid d$$
$$D \rightarrow dB \mid dD.$$

We represent G using techniques from Chapter 4, as the FSR in Figure 12.8. We then represent this as a structure known as an *adjacency matrix*, shown in Table 12.6. The matrix is indexed by node names. The existence of an arc from any node, X, to a node Y is indicated by a 1 in position (X, Y) of the matrix. Note that the matrix we are using does not represent the labels on the arcs of our machine. It could do, if we wished (we simply use the arc label instead of a 1 to indicate an arc), but we do not need this in the discussion that follows.

We now apply Warshall's algorithm (Table 12.7) to the matrix in Table 12.6. The algorithm results in the matrix in Table 12.8, which is known as a *connectivity matrix*. For any two nodes X and Y, in the connectivity matrix, if position (X, Y) is marked with a 1, then there is a path from X to Y in the represented graph.

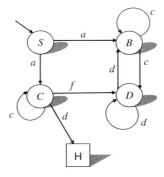

Figure 12.8 A *finite state recogniser* (see Chapter 4).

Table 12.6 An *adjacency matrix*, M, for the finite state recogniser in Figure 12.8, indexed by state names. If $M(X, Y) = 1$, there is an arc from X to Y in the represented machine.

	S	B	C	D	H
S	0	1	1	0	0
B	0	1	0	1	0
C	0	0	1	1	1
D	0	1	0	1	0
H	0	0	0	0	0

Table 12.7 Warshall's algorithm for producing a *connectivity matrix* from an $n \times n$ *adjacency matrix*, M, representing a directed graph. The connectivity matrix tells us which nodes in the graph can be reached from each of the nodes.

```
for k := 1 to n do
  for i := 1 to n do
    for j := 1 to n do
      if M[i, j] = 0 then
          if M[i, k] = 1 and M[k, j] = 1 then
            M[i, j] = 1
          endif
      endif
    endfor
  endfor
endfor
```

For example, the existence of a path from S to D in the FSR (Figure 12.8) is indicated by the 1 in position (S, D) of the matrix. Note that there is no corresponding path from D to S, hence 0 is the entry in position (D, S).

Warshall's algorithm is clearly $O(n^3)$, where n is the number of states in the FSR, since it contains an outer n-loop, encompassing another n-loop which, in turn, encompasses a further n-loop.

Let us return to our problem. We are interested in any state (apart from H) from which there is not a path to the halt state (such states are useless, since

Table 12.8 The *connectivity matrix*, C, derived by applying Warshall's algorithm (Table 12.7) to the adjacency matrix of Table 12.6. If $C(X, Y) = 1$, there is a path (i.e., series of one or more arcs) from X to Y in the represented machine (Figure 12.8). Note that the matrix tells us that the machine cannot reach its halt state from states B or D.

	S	B	C	D	H
S	0	1	1	1	1
B	0	1	0	1	0
C	0	1	1	1	1
D	0	1	0	1	0
H	0	0	0	0	0

they cannot feature in the acceptance of any strings). In Table 12.8, the row for any such state will have a 0 in the column corresponding to the halt state. In Table 12.8, these states are B and D. They serve no purpose and can be removed from the FSR along with all of their ingoing arcs. This results in the FSR shown in Figure 12.9. The corresponding productions can, of course, be removed from the original grammar, leaving:

$$S \to aC$$
$$C \to cC \mid d.$$

The purpose of the above has been to illustrate how an algorithm that runs in time $O(n^3)$ can help us solve a problem that is relevant to Part 1 of this book. There are other algorithms that can achieve the task of removing what are known as *useless* productions from regular and context free grammars. We will not consider them here. There are books in the Further Reading section that discuss such algorithms. There is an exercise based on the approach above at the end of this chapter.

12.5.5 Exponential Time

Now, if the dominant expression in an equation denoting the running time of a program is $O(x^n)$, where $x > 1$ and, again, n is some measure of the size of the input, the amount of execution time rises dramatically as the input length increases. Programs like this are said to take *exponential time* in their execution.

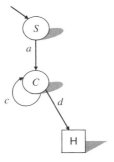

Figure 12.9 The finite state recogniser of Figure 12.8 with its useless states and associated arcs omitted.

Let us briefly consider the *subset algorithm* (Table 4.6 of Chapter 4) that removes non-determinism from a finite state recogniser, M, leaving a deterministic version of the same machine, M^d. The deterministic machine is such that each of its states represents a distinct subset of the states of M. In fact, in the worst case, we could have to create a new state for every single subset of set of states of M. Consider the FSR in Figure 12.10, which represents a worst case scenario for the subset algorithm. If you apply the subset algorithm to this machine (as you are asked to do in an exercise at the end of this chapter), you should produce 8 states, as there are 8 subsets of the set $\{1, 2, 3\}$ (including the empty set).

In general, for a set of n elements, there are 2^n subsets. Thus, in the worst case, the running time of the subset algorithm is, *at the very least*, $O(2^n)$, where n is the number of states in the original non-deterministic machine. This is an algorithm that can have an *exponential running time*. For example, a worst case FSR with 30 states might require the algorithm to create 1 073 741 824 new states for the deterministic machine. A machine that can count up to 1 000 000 in a second would take around 18 minutes to count up to 1 073 741 824. However, consider a worst case machine with 75 states. Our counting machine would take over 1 197 000 000 *years* to count up to the number (2^{75}) of states that the new FSR might require!

We now return, as promised, to the example of the 4-tape deterministic version of the non-deterministic Turing machine from earlier in this chapter. Simplifying somewhat, we can regard this as taking time $O(nk^n)$, where k is the number of quintuples in the non-deterministic machine, in order to find a solution of length n. We saw the striking difference between the time taken by the non-deterministic parallel machine, which was $O(n)$, i.e. of *linear* relationship to the length of the solution, and that taken by the deterministic version, for even small values of n (3 and 11) and small values of k (3 and 10).

We have considered the difference in the time taken by two machines (a non-deterministic TM and its 4-tape deterministic counterpart) in terms of the length of a given solution. However, the disparity between the time taken by the two machines becomes even greater if we also consider the relationship between the length of the *input* and the length of the solution. For example, suppose the

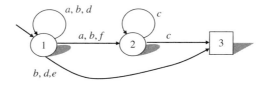

Figure 12.10 A finite state recogniser that represents a worst case situation for the *subset algorithm* of Chapter 4.

non-deterministic machine solves a problem that has minimum length solution n^3, where n is the length of the input. Then the non-deterministic machine will find that solution in time $O(n^3)$. Suppose $n = 11$, then, simplifying things grossly, our deterministic machine will find a solution in the best case in a time $O\left(k^{n^3-1}\right)$, which represents a time of 10^{1330} "moments". Remember, we indicated above that 2^{75} is a pretty big number. Imagine, then, how large 10^{1330} is. The point being made here is that when the non-deterministic machine requires *polynomial* time to find the shortest solution, the deterministic machine constructed from it by our method requires *exponential* time.

12.6 The Implications of Exponential Time Processes

If you find *big O* a little daunting, think about it in this way: for our hypothetical problem, our non-deterministic machine, N, has 11 squares marked on its input tape. Since $n = 11$ and we know that the shortest solution is n^3 instructions in length, N is thus searching for a solution of 11^3 instructions. If N is a parallel machine, as described above, it will find the solution in 11^3 moments.

Now consider the behaviour of D, the deterministic version of N. To get to the first solution of length 11^3, D will have to first generate all sequences of quintuple labels of length $11^3 - 1$. If there are 10 quintuples, there are 10^{11^3-1} distinct sequences of labels of length $11^3 - 1$, i.e. 10^{1330}.

Let us now consider a simpler example, so that we can appreciate the size of the numbers that we are dealing with here. Suppose the shortest sequence of instructions that a particular 10-quintuple non-deterministic machine can find is 101 instructions, for an input of 10 symbols. Before finding the 101-quintuple sequence that yields the solution, the deterministic version will generate all sequences of quintuples of length 100. There are 10^{100} of these sequences, since there are 10 quintuples. The time taken by our deterministic machine is around 10^{100} moments, by the most gross simplification imaginable (e.g., even if we assume that the generation of each sequence takes only a moment, and ignore that our machine will have generated all the sequences of length 99, of length 98, and so on).

At this point we need to buy a faster deterministic machine. We go to the TM store and purchase a machine that does 10^{20} operations in a *year*. This machine does in excess of *three million million operations each second*. If we used it to find the solution to the problem above that took our original deterministic "one-million-instructions-a-second" model 30.5 hours to find, it would take our new machine about $\frac{3}{100}$ of a second! However, our new machine would still take in the order of 10^{80} *years* to solve our new problem, while our non-deterministic machine, at a similar instruction execution rate, would take $\frac{1}{30,000,000,000}$ th of a *second*. Given that, by some estimations, the Earth is at this time around 400 million years old, if we had started off our deterministic machine at the time the Earth was born, it would now be about $\frac{1}{250,000,000,000,000,000,000,000,000,000,000,000,000,000,000,000,000,000,000,000,000,000,000,000}$ th of the way through the time it needs to find the solution.

12.6.1 "Is P Equal to NP?"

From a purely formal point of view, the amount of time taken to solve a problem is not important. The requirements are that a solution be found in a *finite* number of steps, or, put another way in a *finite* period of time. Now, a time period like 10^{80} years may be to us an incredibly long period of time, but nevertheless it is finite. The purpose of the construction we used to create our deterministic machine was to show functional *equivalence*, i.e., that whenever there is a solution that the non-deterministic machine *could* find, the deterministic machine *will* find it (eventually!).

In a more practical vein, in creating a deterministic machine that models one that is non-deterministic, we may also wish to maintain invariance between the two machines in the other dimensions of computational power (*space* and *time*) as well as invariance in *functionality*. Given a non-deterministic TM that finds solutions in polynomial time, we would ideally like to replace it by a deterministic machine that also takes polynomial time (and not *exponential*-time, as is the case for the construction method we have used in this chapter). Clearly, this can be done in many, many cases. For example, it would be possible to model a search program (of the types discussed earlier) deterministically by a polynomial-time TM.

The above discussion leads to a question of formal computer science that is as yet unanswered. The question is:

> For each polynomial-time non-deterministic TM, can we construct an equivalent polynomial-time deterministic TM?

This is known as the "*is P equal to NP?*" problem, where P is the set of all polynomial-time TMs, and NP is the set of all non-deterministic polynomial-time TMs. As yet, no solution to it has been found. Given our analogy between non-deterministic TMs and parallel processing, the lack of a solution to this problem indicates the possibility that there are parallel processes that run in *polynomial* time, that can only be modelled by serial processes that run in *exponential* time. More generally, it suggests that there may be some problems that can be solved in a reasonable amount of time *only* if solved non-deterministically. True non-determinism involves *random* execution of applicable instructions, rather than the systematic execution processes considered above. This means that a truly non-deterministic process is not guaranteed to find a solution even if there is one.

There may therefore be problems that are such that either,

a solution *may* be found in *polynomial* time by a non-deterministic program (but it *may not* actually find a solution even if one exists),

or,

a solution *will* be found in *exponential* time by a deterministic program, if a solution exists, but it may take a prohibitive amount of time to discover it.

12.7 Observations on the Efficiency of Algorithms

We discovered earlier in this part of the book that there are well-defined processes that are deceptively simple in formulation and yet cannot be solved by algorithmic means. In this chapter, we have also discovered that certain processes, despite being *algorithmic* in formulation, may, in certain circumstances be *practically* useless. The caveat "in certain circumstances" is a very important part of the preceding sentence. Algorithms such as the subset algorithm can be very effective, for relatively small sizes of input. In any case, the real value of the subset algorithm is in theoretical terms; its existence demonstrates a property of finite state recognisers. In that sense, discussion of its running time is meaningless. We *know* that it will (eventually) produce a solution, *for every possible assignment of valid input values.* Hence, it represents a general truth, and not a practical solution. The same applies to many of the other algorithms that have supported the arguments in Parts 1 and 2 of this book.

However, as programmers, we know that efficiency is an important consideration in the design of our programs. An optimal algorithm is desirable. Analysis of the running time of algorithms using the big O approach is a useful tool in predicting the efficiency of the algorithm that our program embodies. This chapter has endeavoured to show certain ways in which our knowledge of formal computer science can support and enrich the enterprise of programming. This has hopefully demonstrated, at least in part, the fruitful relationship between the worlds of abstraction and practicality.

EXERCISES

For exercises marked "†", solutions, partial solutions, or hints to get you started appear in "Solutions to Selected Exercises" at the end of the book.

12.1.[†] Justify that the average and worst case running times for the algorithm specified in Figure 12.6 are both $O(n)$.

12.2. Identify the best case scenario for the bubble sort algorithm (Table 12.5). Convince yourself that in such a case, the algorithm has a running time $O(n)$.

12.3. In connection with the above application of Warshall's algorithm to the regular grammar problem:

(a) The solution assumes an adjacency matrix representation of the equivalent finite state recogniser. Sketch out an algorithm to produce an adjacency matrix from a regular grammar, and analyse its running time.

(b) Sketch out the part of the solution that involves searching the final matrix to detect which states do not lie on any path to the halt state, and analyse its running time.
    Note: see the comments for part (c) in "Solutions to Selected Exercises".

(c)[†] The approach being considered here is not guaranteed to identify all useless states. Identify why this is so, and suggest a solution.
    Hint: our method suggests detecting all states which do not lie on a path to a halt state. However, there may be states from which the halt state can be reached that are nevertheless useless.

12.4.[†] Apply the subset algorithm of Chapter 4 (Table 4.6) to the finite state recogniser in Figure 12.10.

12.5. Design an algorithm to carry out the *partitioning* of an array as required by the *quicksort* sort routine (see Figure 12.4). A simple way to choose the *pivot* is to use the first element in the array. The running time of your algorithm should be $O(n)$. Convince yourself that this is the case.

Further Reading

The following are some suggested titles for further reading. Notes accompany most of the items. Some of the titles refer to articles that describe practical applications of concepts from this book.

The numbers in parentheses in the notes refer to chapters in this book.

Church A. (1936) An Unsolvable Problem of Elementary Number Theory. American Journal of Mathematics, 58, 345–363.

> Church and Turing were contemporaneously addressing the same problems by different, but equivalent means. Hence, in books such as Harel's we find references to the "Church–Turing thesis", rather than "Turing's thesis" (9, 10, 11).

Cohen D.I.A. (1996) Introduction to Computer Thery. John Wiley, New York, 2nd edition.

> Covers some additional material such as regular expressions, Moore and Mealy machines (in this book our FSTs are Mealy machines) (8). Discusses relationship between multi-stack PDRs (5) and TMs.

Floyd R.W. and Beigel R. (1994) The Language of Machines: an Introduction to Computability and Formal Languages. W.H. Freeman, New York.

> Includes a discussion of regular expressions (4), as used in the UNIX™ utility "egrep". Good example of formal treatment of minimisation of FSRs (using equivalence classes) (4).

Harel D. (1992) Algorithmics: the Spirit of Computing. Addison-Wesley, Reading, MA, 2 nd edition.

Study of algorithms and their properties, such as complexity, big O running time (12) and decidability (11). Discusses application of finite state machines to modelling simple systems (8). Focuses on "counter programs": simple programs in a hypothetical programming language.

Harrison M.A. (1978) Introduction to Formal Language Theory. Addison-Wesley, Reading, MA.

Many formal proofs and theorems. Contains much on closure properties of languages (6).

Hopcroft J.E. and Ullman J.D. (1979) Introduction to Automata Theory, Languages and Computation. Addison-Wesley, Reading, MA.

Discusses linear bounded TMs for context sensitive languages (11).

Jensen K. and Wirth N. (1975) Pascal User Manual and Report. Springer-Verlag, New York.

Contains the BNF and Syntax chart descriptions of the Pascal syntax (2). Also contains notes referring to the ambiguity in the "if" statement (3).

Kain R.Y. (1972) Automata Theory: Machines and Languages. McGraw-Hill, New York.

Formal treatment. Develops Turing machines before going on to the other abstract machines. Discusses non-standard PDRs (5) applied to context sensitive languages.

Kelley D. (1998) Automata and Formal languages: an Introduction. Prentice Hall, London.

Covers most of the introductory material on regular (4) and context free (5) languages, also has chapters on Turing machine language processing (7), decidability (11) and computational complexity (12).

Minsky M.L. (1967) Computation: Finite and Infinite Machines. Prentice Hall, Englewood Cliffs, NJ.

A classic text, devoted to an investigation into effective procedures (11). Very detailed on most aspects of computer science. Of particular relevance is description of Shannon's 2-state TM result (12), and reference to unsolvable problems (11). The proof we use in this book to show that FSTs cannot perform arbitrary multiplication (8) is based on Minsky's.

Post E. (1936) Finite Combinatory Processes – Formulation 1. Journal of Symbolic Logic, 1, 103–105.

Post formulated a simple abstract string manipulation machine at the same time as did Turing (9). Cohen (see above) devotes a chapter to these "Post" machines.

Rayward-Smith V.J. (1983) A First Course in Formal Language Theory. Blackwell, Oxford, UK.

The notation and terminology for formal languages we use in this book is based on Rayward-Smith. Very formal treatment of regular languages (plus regular expressions), FSRs (4), and context free languages and PDRs (5). Includes Greibach normal form (as does Floyd and Beigel) an alternative CFG manipulation process to Chomsky Normal Form (5). Much material on top-down and bottom-up parsing (3), LL and LR grammars (5), but treatment very formal.

Rich E. and Knight K. (1991) Artificial Intelligence. McGraw-Hill, New York.

Artificial intelligence makes much use of representations such as grammars and abstract machines. In particular, machines called recursive transition networks and augmented transition networks (equivalent to TMs) are used in natural language processing.

Tanenbaum A.S. (1998) Computer Networks. Prentice-Hall, London. 3rd edition.

Discusses FSTs (8) for modelling protocol machines (sender or receiver systems in computer networks).

Turing A. (1936) On Computable Numbers with an Application to the Entscheidungsproblem. Proceedings of the London Mathematical Society, 42, 230–265.

The paper in which Turing introduces his abstract machine, in terms of computable numbers rather than computable functions. Also includes his notion of a universal machine (10). A paper of remarkable contemporary applicability, considering that Turing was considering the human as computer, and not machines.

Winston P.H. (1992) Artificial Intelligence. Addison-Wesley, Reading, MA (see Rich), 3rd edition.

Wood D. (1987) Theory of Computation. John Wiley, Chichester, UK.

Describes several extensions to PDRs (5). Introduction to proof methods, including the pigeonhole principle (also mentioned by Harel) on which both the repeat state theorem (6, 8) and the *uvwxy* theorem (6) are based.

Solutions to Selected Exercises

Chapter 2

1. (a) *Regular.* Every production has a lone non-terminal on its left-hand side, and the right-hand sides consist of either a single terminal, or a single terminal followed by a single non-terminal.

   (b) *Context free.* Every production has a lone non-terminal on its left-hand side. The right-hand sides consist of arbitrary mixtures of terminals and/or non-terminals, one of which ($aAbb$) does not conform to the pattern for regular right-hand sides.

2. (a) $\{a^{2i}b^{3j} : i, j \geq 1\}$ i.e., as followed by bs, any number of as divisible by 2, any number of bs divisible by 3

   (b) $\{a^i a^j b^{2j} : i \geq 0, j \geq 1\}$ i.e., zero or more as, followed by one or more as followed by twice as many bs

   (c) $\{\}$ i.e., the grammar generates no strings at all, as no derivations beginning with S produce a terminal string

   (d) $\{\varepsilon\}$ i.e., the only string generated is the empty string.

3. xyz, where $x \in (N \cup T)^*$, $y \in N$, and $z \in (N \cup T)^*$
   The above translates into: "a possibly empty string of terminals and/or non-terminals, followed by a single non-terminal, followed by another possibly empty string of terminals and/or non-terminals".

5. For an alphabet A, A^* is the set of all strings that can be taken from A including the empty string, ε. A regular grammar to generate, say $\{a, b\}^*$ is

$$S \rightarrow \varepsilon \mid aS \mid bS.$$

ε is derived directly from S. Alternatively, we can derive a or b followed by a or b or ε (this last case terminates the derivation), the a or b from the last stage being followed by a or b or ε (last case again terminates the derivation), and so on ...

Generally, for any alphabet, $\{a_1, a_2, \dots, a_n\}$ the grammar

$$S \rightarrow \varepsilon \mid a_1 S \mid a_2 S \mid \ \dots \ \mid a_n S$$

is regular and can be similarly argued to generate $\{a_1, a_2, \dots, a_n\}^*$.

7. (b) The following fragment of the BNF definition for Pascal, taken from Jensen and Wirth (1975), actually defines all Pascal expressions, not only Boolean expressions.

        \<expression> ::= \<simple expression> | \<simple expression>
            \<relational operator> \<simple expression>
        \<simple expression> ::= \<term> | \<sign> \<term> |
            \<simple expression> \<adding operator> \<term>
        \<adding operator> ::= + | – | or
        \<term> ::= \<factor> | \<term> \<multiplying operator> \<factor>
        \<multiplying operator> ::= * | / | div | mod | and
        \<factor> ::= \<variable> | \<unsigned constant> | (\<expression>) |
            \<function designator> | \<set> | not \<factor>
        \<unsigned constant> ::= \<unsigned number> | \<string> |
        \<constant identifier> | nil
        \<function designator> ::= \<function identifier> |
            \<function identifier> (\<actual parameter>
                      {, \<actual parameter})
        \<function identifier> ::= \<identifier>
        \<variable> ::= \<identifier>
        \<set > ::= [\<element list>]
        \<element list> ::= \<element> {, \<element> } | \<empty>
        \<empty> ::=

Note the use of the empty string to enable an empty \<set> to be specified (see Chapter 5).

10. Given a finite set of strings, each string, x, from the set can be generated by regular grammar productions as follows:

$$x = x_1 x_2 \ldots x_n, \; n \geq 1$$
$$S \rightarrow x_1 X_1$$
$$X_1 \rightarrow x_2 X_2$$
$$.$$
$$.$$
$$X_{n-1} \rightarrow x_n$$

For each string, x, we make sure the non-terminals X_i, are unique (to avoid any derivations getting "crossed").

This applies to any finite set of strings, so any finite set of strings is a regular language.

Chapter 3

1. (b) $\{a^i b^j c^k : i, j, k \geq 1, \, i = j \; or \; j = k\}$

i.e., strings of as followed by bs followed by cs, where the number of as equals the number of bs, or the number of bs equals the number of cs, or both.

(c) Grammar G can be used to draw two different derivation trees for the sentence $a^3 b^3 c^3$, as shown in Figure S.1.

The grammar is thus ambiguous.

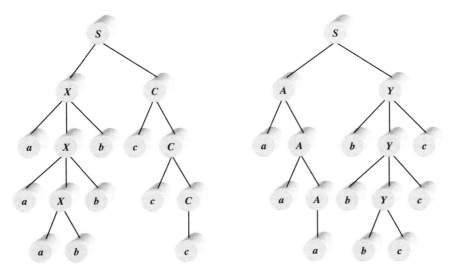

Figure S.1 Two derivation trees for the same sentence $(a^3 b^3 c^3)$.

2. (b) As for the Pascal example, the semantic implications should be discussed in terms of demonstrating that the same statement yields different results according to which derivation tree is chosen to represent its structure.

Chapter 4

1. (a) The FSR obtained directly from the productions of the grammar is shown in Figure S.2.

   The FSR in Figure S.2 is non-deterministic, since it contains states (for example, state A) with more than one identically labelled outgoing arc going to different destinations.
   The deterministic version, derived using the subset method (null state removed) is in Figure S.3.

3. One possibility is to represent the FSR as a two-dimensional table (array), indexed according to (state, symbol) pairs. Table S.1 represents the FSR of exercise 1(a) (Figure S.2).
   In Table S.1, element (A, a), for example, represents the set of states ($\{S, B\}$) that can be directly reached from state A given the terminal a. Such a representation would be easy to create from the productions of a grammar that could be entered by the user, for example. The representation is also highly useful for creating the deterministic version. This version is also made more suitable if the language permits dynamic arrays (Pascal does not, but C and Java are languages that do). The program also needs to keep details of which states are halt and start states.

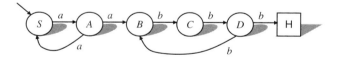

Figure S.2 A non-deterministic finite state recogniser.

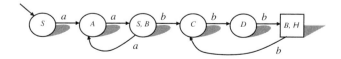

Figure S.3 A deterministic version of the FSR in Figure S.2.

Table S.1 A tabular representation of
the finite state recogniser in Figure S.2.

States	Terminal symbols	
	a	*b*
S	*A*	–
A	*S, B*	–
B	–	*C*
C	–	*D*
D	–	*B, H*
H	–	–

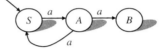

Figure S.4 Part of the finite state recogniser from Figure S.2.

Figure S.5 The FSR fragment from Figure S.4 represented as a linked list of
(state, arc symbol, next state) triples.

An alternative scheme represents the FSR as a list of triples, each triple
representing one arc in the machine. In languages such as Pascal or C, this
scheme can be implemented in linked list form. For example, consider the part
of the FSR from exercise 1 (Figure S.2) shown in Figure S.4.

This can be represented as depicted in Figure S.5.

This representation is particularly useful when applying the reverse operation in the minimisation algorithm (the program simply exchanges the first
and third elements in each triple). It is also a suitable representation for
languages such as LISP or PROLOG, where the list of triples becomes a list
of three element lists.

Since FSRs can be of arbitrary size, a true solution to the problem of
defining an appropriate data structure would require dynamic data structures, even down to allowing an unlimited source of names. Although this is

probably taking things to extremes, you should be aware when you are making such restrictions, and what their implications are.

Chapter 5

2. First of all, consider that a DPDR is not necessarily restricted to having one halt state. You design a machine, M, that enters a halt state after reading the first a, then remains in that state while reading any more as (pushing them on to the stack, to compare with the bs, if there are any). If there are no bs, M simply stops in that halt state, otherwise on reading the first b (and popping off an a) it makes a transition to another state where it can read only bs. The rest of M is an exact copy of $M_3{}^d$, of Chapter 5. If there were no bs, any as on the stack must remain there, even though M is accepting the string. Why can we not ensure that in this situation, M clears the as from its stack, but is still deterministic?

5. The reasons are similar to why arbitrary palindromic languages are not deterministic. When the machine reads the as and bs part of the string, it has no way of telling if the string it is reading is of the "number of as = number of bs", or the "number of bs = number of cs" type. It thus has to assume that the input string is of the former type, and backtrack to abandon this assumption if the string is not.

6. One possibility is to represent the PDR in a similar way to the list representation of the FSR described above (sample answer to exercise 3, Chapter 4). In the case of the PDR, the "current state, input symbol, next state" triples would become "quintuples" of the form:

   "current state, input sym, pop sym, push string, new state".

   The program could read in a description of a PDR as a list (a file perhaps) of such "rules", along with details of which states were start and halt states.

   The program would need an appropriate dynamic data structure (e.g. linked list) to represent a stack. It may therefore be useful to design the stack and its operations first, as a separate exercise.

   Having stored the rules, the program would then execute the algorithm in Table S.2.

   As the PDR is deterministic, the program can assume that only one quintuple will be applicable at any stage, and can also halt its processing of

Table S.2 An algorithm to simulate the behaviour of a deterministic pushdown recogniser. The PDR is represented as quintuples.

---

can-go := true
C := the start state of the PDR

while not(end-of-input) and can-go
    if there is a quintuple, Q, such that
        Q's current state = C, and
        Q's pop sym = the current symbol "on top" of the stack, and
        Q's read sym = the next symbol in the input
    then
        remove the top symbol from the stack
        set up ready to read the next symbol in the input
        push Q's push string onto the stack
        set C to be Q's next state
    else
        can-go := false
    endif
endwhile

if end-of-input and C is a halt state then
    return("yes")
else
    return("no")
endif

---

invalid strings as soon as an applicable quintuple cannot be found. The non-deterministic machine is much more complex to model: I leave it to you to consider the details.

Chapter 6

3. Any FSR that has a loop on some path linking its start and halt states *in which there is one or more arcs not labelled with* ε recognises an infinite language.

4. In both cases, it is clear that the *v* part of the *uvw* form can consist only of *a*s or *b*s or *c*s. If this were not the case, we would end up with symbols out of their respective correct order. Then one simply argues that when the *v* is repeated the required numeric relationship between *a*s, *b*s and *c*s is not maintained.

5. The language specified in this case is the set of all strings consisting of two copies of any string of *a*s and/or *b*s. To prove that it is not a CFL, it is useful to use the fact that we know we can find a *uvwxy* form for which $|vwx| \leq 2^n$, and n is the number of non-terminals in a Chomsky Normal Form grammar to generate our language. Let $k = 2^n$ There are many sentences in our language of the form $a^n b^n a^n b^n$, where $n > k$ Consider the *vwx* form as described immediately above. I leave it to you to complete the proof.

Chapter 7

1. This can be done by adding an arc labelled x/x (N) for each symbol, x, in the alphabet of the particular machine (including the blank) from each of the halt states of the machine to a new halt state. The original halt states are then designated as non-halt states. The new machine reaches its single halt state leaving the tape/head configuration exactly as did the original machine.

Chapter 8

1. (The "or" FST) Assuming that the two input binary strings are the same length and are interleaved on the tape, a bitwise *or* FST is shown in Figure S.6.

2. An appropriate FST is depicted in Figure S.7.

Chapter 9

4. Functions are discussed in more detail in Chapter 11. A specification of a function describes what is computed and does not go into detail about how the computation is done. In this case, then, the TM computes the function:

$$f(y) = y \ div \ 2, \ y \geq 1.$$

5. (a) Tape on entry to loop: $d1^{x+1}e$
    Tape on exit: $df^{x+1}e1^{x+1}$ i.e., the machine copies the $x+1$ 1s between d and e to the right of e, replacing the original 1s by fs.

   (b) $f(x) = 2x + 3$

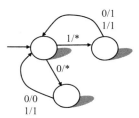

Figure S.6 A bitwise "or" finite state transducer. The two binary numbers are the same length, and interleaved when presented to the machine. The machine outputs "*" on the first digit of each pair.

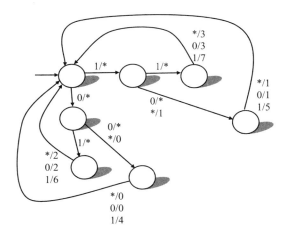

Figure S.7 A finite state transducer that converts a binary number into an octal number. The input number is presented to the machine in reverse, and terminated with "*". The answer is also output in reverse.

Chapter 10

3. The sextuple $(1, a, A, R, 2, 2)$ of the three-tape machine M might be represented as shown in Figure S.8.

Chapter 11

1. Assume the TM, P, solves the printing problem by halting with output 1 or 0 according to whether M would, or would not, write the symbol s.

   P would need to be able to solve the halting problem, or in cases where M was not going to halt P would not be able to write a 0 for "no".

Chapter 12

1. There are two worst case scenarios. One is that the element we are looking for is in position $(n, 1)$, i.e., the leftmost element of the last row. The other is that the element is not in the array at all, but is greater than every element at the end of a row except the one at the end of the last row, and smaller than every element on the last row. In both of these cases we inspect every element at the

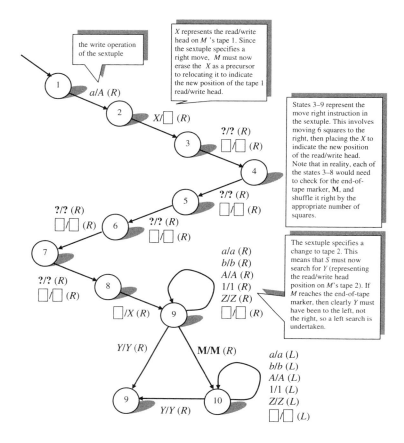

Figure S.8 A sketch of how a single-tape TM, S, could model the sextuple $(1, a, A, R, 2, 2)$ of a 3-tape machine, M, from Chapter 10. For how the three tapes are coded onto S's single tape, see Figures 10.21 and 10.22. This sequence of states applies when S is modelling state 1 of M and S's read/write head is on the symbol representing the current symbol on M's tape 1.

end of a row (there are n of these), then every element on the last row (there are $n-1$ of these, as we already inspected the last one). Thus, we inspect $2n-1$ elements. This represents time $O(n)$.

There are also two average case scenarios. One is that the element is found midway along the middle row. The other is that the element is not in the array at all, but is greater than every element at the end of a row above the middle row, smaller than every element in the second half of the middle row and greater than the element to the left of the middle element in the middle row. In this case, we inspect half of the elements at the end of the rows (there are $n/2$ of

Table S.3 The deterministic version of the finite state recogniser from Figure 12.10, as produced by the subset algorithm of Chapter 4 (Table 4.6). There are 8 states, which is the maximum number of states that can be created by the algorithm from a 3-state machine.

States	Terminal symbols					
	a	b	c	d	e	f
1 (start state)	1_2	1_2_3	N	1_3	3	2
2	N	N	2_3	N	N	N
3 (halt state)	N	N	N	N	N	N
1_2	1_2	1_2_3	2_3	1_3	3	2
1_3 (halt state)	1_2	1_2_3	N	1_3	3	2
2_3 (halt state)	N	N	2_3	N	N	N
1_2_3 (halt state)	1_2	1_2_3	2_3	1_3	3	2
N (null state)	N	N	N	N	N	N

these), and we then inspect half of the elements on the middle row (there are $n/2-1$) of these, since we already inspected the element at the end of the middle row. We thus make $n/2 + n/2 - 1 = n- 1$ comparisons.

This, once again, is $O(n)$.

3. (c) A further form of useless state, apart from those from which the halt state cannot be reached, is one that cannot be reached from the start state. To find these, we simply examine the row in the connectivity matrix for the start state, S, of the machine. Any entry on that row (apart from position S, S) that is not a 1 indicates a state that cannot be reached from S, and is thus useless. If there are n states in the machine, this operation requires time $O(n)$.

With respect to part (b) of the question, the *column* for a state indicates the states from which that state can be reached. Entries that are not 1 in the column for the halt state(s) indicate states from which the halt state cannot be reached (except, of course for entry H, H where H is the halt state in question). This operation is $O(m \times n)$ where n is the total number of states, and m is the number of halt states. The running time is thus never worse that $O(n^2)$ which would be the case if all states were halt states. For machines with a single halt state it is, of course $O(n)$.

4. Table S.3 shows the result of applying the subset algorithm (Table 4.6) to the finite state recogniser of Figure 12.10. For an example see the finite state recogniser in Figure S.2, which is represented in tabular form in Table S.1. Here, the tabular form is used in preference to a diagram, as the machine has a rather complex structure when rendered pictorially.

Index